Science Learning

Processes and Applications

Carol Minnick Santa
School District #5, Kalispell, Montana

Donna E. Alvermann
University of Georgia, Athens, Georgia

Editors

INTERNATIONAL READING ASSOCIATION
800 Barksdale Road
Newark, Delaware 19714-8139

The International Reading Association attempts, through its publications, to provide a forum for a wide spectrum of opinions on reading. This policy permits divergent viewpoints without assuming the endorsement of the Association.

Library of Congress Cataloging in Publication Data

Science Learning: processes and applications/Carol Minnick Santa,
Donna E. Alvermann, editors.
p. cm.
Includes bibliographical references and index.
1. Science—Study and teaching (Elementary) 2. Science—
study and teaching (Secondary) 3. Learning. 4. Reading.
I. Santa, Carol Minnick. II. Alvermann, Donna E.
Q181.S3745 1991 91-2473
372.3'5044—dc20 CIP
ISBN 0-87207-366-1

Graphic Design: Larry Husfelt
Typesetting: Wendy Mazur
Staff Editors: Romayne McElhaney, Karen Goldsmith, John Micklos
Cover Photo Credits:
Mary Loewenstein-Anderson, science classroom
Larry Husfelt, sugar maple

Contents

Foreword

Science Learning: Processes and Applications contains an excellent collection of articles authored by science and reading teachers at all levels. It is a combination of the "why" and "how" of learning from science texts and materials. The authors have worked together toward advancing the goal of scientific literacy by incorporating into their work the current thinking and research on effective teaching.

Scientific literacy has been a major concern of educators for a number of years. My monograph, *Improving Reading in Science* (last revised in 1984), is a part of IRA's popular Reading Aids Series. The interest shown in my small book indicated the need for an expanded and current publication.

The need for such a work was underscored in 1985, when the American Association for the Advancement of Science (AAAS) began Project 2061 (named for the year when Halley's comet will return). The project, also called Science for All Americans, recommends that all students have a knowledge of science, mathematics, and technology because that knowledge makes the world more comprehensible and more interesting.

This volume will be of value to anyone who endorses that world vision and will make an excellent addition to the outstanding collection of publications already provided by IRA.

Judith N. Thelen
Frostburg State University

Introduction

As teachers we constantly seek ways to nourish our professional lives. Teaching is often lonely. We tend to sequester ourselves in classrooms and discrete departments. Seldom do we collaborate with those outside of our areas of expertise. Yet knowledge cannot be departmentalized into content categories. We need to stretch across grade levels and discipline boundaries to examine teaching from multiple points of view.

This book is a beginning. It reflects collaboration in terms of content areas, levels, and audience. Some of the authors are science teachers; others are from the field of reading. Some teach in public schools, others in universities. Some chapters address elementary and secondary classroom teachers; others address teacher educators interested in helping students learn science. In this volume, science teachers and reading teachers seek to understand and help one another fine tune their craft.

A quick look at the table of contents reveals a variety of topics. Some chapters examine science and reading along theoretical dimensions, while others focus on practical teaching ideas. Although the focus of each chapter is somewhat different, common themes bind this work into a cohesive whole.

For example, science and reading teachers have remarkably similar goals for their students. Foremost is the pursuit of meaning. Scientists assume that our world and our universe can be understood. They believe in causality and the orderliness of the natural world—and the importance of teaching students ways to discover these dimensions. Similarly, the goal of reading is understanding. Both science and reading involve students in making sense of information. Each chapter in this book describes ways to make science learning more meaningful.

The problem with the pursuit of meaning in this field is that science concepts and science texts are often difficult to understand. The authors here recognize this problem and provide suggestions for meeting the challenge it presents. All agree that science and reading teachers must show students how to learn from science texts; the texts are too difficult for students to tackle on their own.

Another common theme deals with science and reading process skills. While the investigative and comprehension processes are somewhat different, processing strategies remain at the heart of both disciplines. Science teachers want students to internalize such processes as observation, experimentation, and communication. They want their students to leave their classrooms with the tools for observing and explaining natural phenomena.

As with science, reading specialists want their students to know strategies for creating meaning from text, including those for developing background knowledge, using text structure, organizing information, and understanding new vocabulary. Each author in this volume focuses on the processes of reading and thinking about science text.

Active learning is another common theme. The most effective learners are actively engaged in learning through observing, reading, and experimenting. Science and reading teachers know that students have difficulty learning content simply by reading or observing. Students must participate in science activities, not just read and answer chapter questions. The authors here focus on ways to help students become actively involved in learning science.

Each author also stresses the need for students to gain control of their own learning. Science teachers want students to be capable of asking their own questions and finding their own solutions. They want students to leave class with the tools to challenge their own ideas about nature, as well as those of others. Reading specialists hold similar goals. They want their students to become independent, strategic readers. When facing difficult texts on their own, students need to know which strategies to use and how to use them to comprehend and learn science content.

Yet each author acknowledges that students do not automatically become independent learners. The road to independence is a gradual process. Teachers first must show students how to learn, and then gradually give them more responsibility as they become capable of assuming it. The teacher assumes the role of a coach, guiding students until they can perform without adult assistance.

Although many commonalities bind science and reading, important differences remain. In the final section, John Guthrie addresses these differences. He emphasizes that learning from a textbook and learning science process skills will always have unique characteristics. Although learning to read science texts and learning to think scientifically share many features, they are not the same. The goal of this volume is to capture both the commonalities and the differing points of view. Such diversity promises exciting opportunities for professional growth.

The bulk of this volume is organized into four parts, with each part containing chapters that focus on similar topics. The parts flow along a continuum moving from a more theoretical to a more applied emphasis. Part 1 takes a close look at science and reading processes. We get an inside view of critical features of science learning and reading comprehension and also gain a conceptual framework for the ideas contained in the more applied sections that follow. Part 2 focuses on two critical issues confronting all reading and science educators: the difficult nature of science texts and students' lack of appropriate background knowledge. The authors examine these issues thoroughly and offer suggestions for classroom practice.

In Part 3, we go inside classrooms to see different approaches teachers take to help students learn from science texts. The authors grapple with such important questions as: What is the role of the text in science classrooms? How should students interact with text to encourage critical thinking in science? Part 4 is filled with practical ideas for instruction. We learn how to teach students to pull main ideas from science lessons and to organize information. We also discover how to help students learn new vocabulary and internalize new concepts through writing and making images of what they read.

The book's final section ties the preceding sections together. All in all, we have a volume in which teachers generously share what they know across disciplines and levels of expertise.

CMS
DEA

Contributors

Donna E. Alvermann
University of Georgia
Athens, Georgia

Bonnie B. Armbruster
Center for the Study of Reading
University of Illinois
Champaign, Illinois

Mark W. Aulls
McGill University
Montreal, Quebec
Canada

Linda Baker
University of Maryland
Baltimore County
Baltimore, Maryland

Audrey B. Champagne
State University of New York
at Albany
Albany, New York

Jane L. Davidson
Northern Illinois University
DeKalb, Illinois

Fred N. Finley
University of Minnesota
Minneapolis, Minnesota

John T. Guthrie
University of Maryland
College Park, Maryland

Shirley Harrison
School District #5
Kalispell, Montana

Lynn T. Havens
School District #5
Kalispell, Montana

Kathleen A. Hinchman
Syracuse University
Syracuse, New York

William G. Holliday
University of Maryland
College Park, Maryland

Leopold E. Klopfer
University of Pittsburgh
Pittsburgh, Pennsylvania

Bonnie C. Konopak
Louisiana State University
Baton Rouge, Louisiana

Linda A. Meyer
Center for the Study of Reading
University of Illinois
Champaign, Illinois

K. Denise Muth
University of Georgia
Athens, Georgia

Nancy D. Padak
Kent State University
Kent, Ohio

Michael J. Padilla
University of Georgia
Athens, Georgia

Rosemary K. Lund Padilla
Oconee County Public Schools
Watkinsville, Georgia

Kathleen J. Roth
Michigan State University
East Lansing, Michigan

Carol Minnick Santa
School District #5
Kalispell, Montana

Barbara J. Walker
Eastern Montana College
Billings, Montana

Paul T. Wilson
Western Michigan University
Kalamazoo, Michigan

PART 1

Reading and Science Learning Processes

The logical place to begin this volume is with a discussion of metacognition and science process skills. Both form the theoretical cores of their respective disciplines. Metacognitive processes are fundamental to reading comprehension, and science process skills are basic to science learning.

The chapters in this section focus on the commonalities and differences between metacognition and science learning processes. In so doing, they introduce the major themes woven throughout the book. These themes reflect the importance of helping students pursue meaning by becoming active learners and taking control of their own learning.

Although theoretical in nature, these chapters set the stage for instructional practice in science and reading. Teacher educators in these fields, as well as classroom teachers, will find the information presented here essential for building their own philosophies of instruction.

CHAPTER 1 *Metacognition, Reading, and Science Education*

Linda Baker

In this chapter, Baker considers the relevance of metacognition for science education at the elementary and secondary school levels. Although many writers have alluded to the connection between these two domains, Baker describes the connection in detail. She begins by defining metacognition and examining how metacognitive skills differ among readers of varying maturity. Next she draws parallels between science process skills and metacognition. She concludes with recommendations for improving science learning by enhancing metacognitive skills.

The concept of metacognition is one of the most important contributions cognitive psychologists have made to the field of reading education. Metacognition refers to the awareness and control individuals have over their cognitive processes. Over the past decade, a growing body of evidence has made clear that metacognition plays an important role in effective comprehension and retention of text (see Baker, 1985b; Baker & Brown, 1984a, 1984b; Brown, Armbruster, & Baker, 1986; Garner, 1987).

Metacognitive skills are applicable not only to reading but also to writing, speaking, listening, studying, problem solving, and any other domain requiring cognitive processes. Understanding metacognition can help science educators provide better ways for their students to learn from text materials. In addition, it can help them foster independence in learning through lectures, discussion, laboratory work, and hands-on activities.

Metacognition and Reading

What Is Metacognition?

The term metacognition means different things to different people, but most agree that it involves two basic components: knowledge about cognition and regulation of cognition. The first component concerns the ability to reflect on our own cognitive processes, and includes knowledge about when, how, and why to engage in various cognitive activities. Metacognitive knowledge encompasses our characteristics as learners, the aspects of a cognitive task, and the use of strategies for completing such a task (Flavell, 1981). We may have the personal knowledge that we read more slowly than most people, the task knowledge that it is harder to remember a story for its details than for its general ideas, and the strategy knowledge that a good way to determine the meaning of an unknown word is to look at the surrounding context.

Metacognitive regulation concerns the use of strategies that enable us to control our cognitive efforts. These strategies include planning our moves, checking the outcomes of our efforts, evaluating the effectiveness of our actions and remediating any difficulties, and testing and revising our learning techniques. These general problem solving strategies can be applied in many different situations. However, the implementation of the strategies may differ depending on whether the goal is studying for a test, writing an essay, solving an algebra problem, or performing a science experiment.

One of the most important self-regulatory skills for reading is monitoring comprehension, which involves deciding whether we have understood (evaluation) and taking appropriate steps to correct whatever comprehension problems are noted (regulation). This aspect of metacognition will be a major focus of the chapter.

The distinction between metacognitive strategies and cognitive strategies is somewhat fuzzy. Many study or comprehension strategies traditionally regarded as cognitive are valuable not only because they improve learning or comprehension but also because they provide students with a means of gauging the success of their efforts to learn or comprehend. Such strategies include identifying important information, relating new information to prior knowledge, generating questions, making predictions, and producing summaries.

For practical purposes, it does not matter whether a strategy is labeled cognitive or metacognitive, as long as it is effective. For example, identifying the main idea of a passage is a crucial comprehension skill. Thus, an instructional strategy to foster main idea identification can be considered a cognitive strategy. Identifying the main idea also is an effective way to test comprehension, and thus can be considered a metacognitive strategy.

The primary purpose of providing students with instruction in metacognition is to enable them to take responsibility for their own learning and comprehension activities. Psychologists believe that the best way to achieve this goal is by gradually transferring responsibility for regulation from more knowledgeable persons to the youngsters (Vygotsky, 1978). According to Vygotsky, children first learn how to engage in various problem solving tasks, including reading, through social interaction with others (usually parents or teachers). Initially the adult takes responsibility for regulating the child's activity by setting goals, planning, evaluating, and focusing attention on what is relevant. The adult relinquishes responsibility little by little as the child becomes capable of assuming it, until finally the child internalizes the regulatory mechanisms and can perform without adult assistance.

How Is Metacognition Studied?

A variety of research methods have been used to study metacognition. Much of what we know about metacognitive knowledge of reading comes from students' responses to interview questions such as: "What is the purpose of reading? If you could read only a few sentences in a story, which ones would you read? What do you do if you come to a word you don't understand? What do you do if you come to a sentence you don't understand?" (Paris & Jacobs, 1984).

Much of what we know about comprehension monitoring comes from students' reactions to passages that are difficult to understand. To the extent that students keep track of testing their understanding effectively—using relevant criteria or standards of evaluation—they should be aware of comprehension problems. Asking students to think aloud and collecting process measures of ongoing reading behavior provide information about students' evaluation and regulation of their comprehension and learning. (See Baker & Brown, 1984a, and Garner, 1987, for further discussion of research methods.)

How Do Metacognitive Skills Differ Among Students?

There are differences in knowledge and regulation of cognition between older and younger children and between skilled and less skilled readers of the same age. These differences include views of reading as well as use of strategies for learning and evaluating learning.

Concept of reading. Older and better readers are more likely to regard reading as a meaning-getting process, whereas younger and poorer readers view it as a decoding process (Baker, 1989; Garner, 1981; Paris & Jacobs, 1984; Paris & Myers, 1981).

Purposes for reading. Older and better readers recognize that strategies should change to match the goals of the task (Forrest-Pressley & Waller, 1984).

Strategies for evaluating understanding. Older and better readers recognize the value of paraphrasing or summarizing as a way of testing their understanding. They realize that the ability to remember often reflects their level of understanding (Brown & Smiley, 1977).

Standards for evaluating comprehension. The choice of evaluation standards or criteria depends on knowing what kinds of things make text hard to understand—difficult vocabulary, poor text structure, and lack of correspondence with prior knowledge. Older and better readers use multiple standards for checking their comprehension.

Strategies for dealing with comprehension failures. Older and better readers use a greater variety of fix-up strategies, such as rereading, looking ahead, using surrounding context, trying to make an inference to resolve the problem, and consulting an outside source. Those younger and poorer readers who use fix-up strategies tend to focus their efforts on word-level problems.

Strategies for promoting learning and retention of text. Younger and poorer readers are likely to believe that reading and rereading are sufficient study strategies, whereas older and better readers more often recognize the need to take an active role. They know the value of generating questions for self testing and are better able to construct questions that involve the main ideas of passages rather than trivial information (Andre & Anderson, 1979; Palincsar & Brown, 1984). They also know the value of summarization, recognizing that it is useful as a strategy to promote comprehension, to monitor comprehension, and to promote retention of material for a test (Brown & Day, 1983; Garner, 1985).

How Do We Know When We Don't Understand?

As noted earlier, in order to decide whether we understand, we need to adopt and apply appropriate standards of evaluation. Baker (1985b) recently identified six such standards used by proficient readers:

1. The lexical standard involves checking that the meaning of each word is understood. Given the important role vocabulary knowledge plays in comprehension, use of this standard is vital.
2. The external consistency standard involves checking that the ideas in the text are true or plausible with respect to what we already know. Evaluating the truthfulness of text traditionally is regarded as a component of critical reading, but given the importance of activating prior knowledge when reading, the standard is essential for general comprehension.
3. The propositional cohesiveness standard involves checking that the relationship among propositions (ideas) that share a local context is cohesive. When we read or listen to a passage, we search for connections that allow us to link ideas. Pronouns are one type of tie writers use for this linkage; ambiguous

pronouns can disrupt propositional cohesiveness.

4. The structural cohesiveness standard involves checking to see whether ideas in a text or paragraph are thematically compatible. Textbooks frequently contain problems of structural cohesiveness. It is not unusual to read a paragraph that has no clear main idea, either explicit or implicit, or that includes irrelevant information.
5. The internal consistency standard involves checking the ideas expressed in the text for logical consistency. This task frequently requires the reader to integrate information scattered throughout the text. Evaluation of internal consistency also is traditionally regarded as an important component of critical reading.
6. The informational completeness standard involves checking that the text contains all the information necessary to accomplish a specific goal. Applying this standard is particularly important when reading procedural text, such as instructions for assembling objects or performing experiments. The standard also plays a role in comprehending expository text. It is important to recognize a text that does not include sufficient information to explain a particular point.

It should come as no surprise that students use few of these standards spontaneously. In fact, younger and poorer readers frequently rely exclusively on the lexical standard: they decide they understand a passage if they understand the meaning of each word (Baker, 1984a, 1989). It is encouraging, however, that students who do not use particular standards spontaneously often are quite capable of doing so with instruction. This is true for children as young as age 5 years as well as for adults (Baker, 1984a, 1984b, 1985a; Baker & Zimlin, 1989).

Consider an instructional study Baker and Zimlin conducted with fourth graders. The experimenter first modeled use of a subset of the standards, then gave the children practice and feedback applying them independently. After training, students read a narrative passage requiring the use of all six standards and were asked to identify sections of the text that were hard to understand. A control group of children who did not receive any instruction in using these standards was tested on the same narrative text.

On the test, children in the instructed groups used more standards, and used them more effectively, than those in the control group. Perhaps more important, the instructed group maintained this advantage 2 to 3 weeks after the initial training and for standards other than those in which they were trained. Even less skilled readers benefited from the instruction. The children's use of evaluation standards other than those they were explicitly instructed to use is particularly important from an education standpoint. It would be difficult (if not impossible) to teach children about all possible situations where comprehension difficulties might arise.

Implications for the Classroom

Research in metacognition has consistently found the same pattern of results: older and better readers demonstrate more knowledge and control of their own cognitive processes. Although there is a strong relationship between metacognitive skill and reading achievement, it would be erroneous to conclude that the metacognitive skills of older and better readers are fully developed. Even at the high school and college levels, students frequently fail to monitor their comprehension effectively (Baker, 1985a; Glenberg, Wilkinson, & Epstein, 1987; Otero, 1987). This is not surprising, given that they probably never received direct instruction in using metacognitive strategies. Classroom practices are changing, but the consensus remains that most teachers seldom engage in such instruction (Durkin, 1984).

Good readers pick up metacognitive skills on their own, but poorer readers frequently do not. Given the apparent link between metacognition and achievement, it makes little sense to leave the acquisition of metacognitive skills to chance. Rather, teachers should give students direct instruction in using metacognitive skills from the early elementary years. Although we do not yet have a solid body of research supporting this recommendation, several recent studies offer promising evidence (Baker & Zimlin, 1989; Elliott-Faust & Pressley, 1986; Miller, 1985; Palincsar & Brown, 1984; Paris, Cross, & Lipson, 1984).

Metacognition and Science Education

The literature on metacognition has much to offer the science educator interested in improving students' comprehension and retention of science concepts. The following quotation illustrates the common goals of those in each of these fields: "An essential goal of science education is to encourage students to develop into independent learners—able to acquire information from many sources, to weigh alternatives, and to reach defensible conclusions" (Watson, 1983, p. 62).

Despite the obvious relevance, few authors have pointed out the connection between the literature on metacognition and that on science education. Science educators are aware of the contributions of psychology to their discipline, as evidenced by the considerable attention devoted to the developmental theories of Piaget and Bruner (Collette & Chiapetta, 1984; Esler & Esler, 1985; Peterson et al., 1984; Pope & Gilbert, 1983). However, the information processing or cognitive perspective, from which the concept of metacognition derives, is less frequently acknowledged (Resnick, 1983).

A survey of several science education journals revealed few explicit mentions of metacognition, even though several articles dealt with skills that are unquestionably metacognitive. Most articles that did discuss metacognition were by researchers whose primary affiliation was not in science education. For example, Shuell (1987) relates the findings of cognitive psychology to science education, acknowledging the importance of metacognition. DiSessa (1987) explicitly discusses the importance of strategic and metacognitive knowledge in science and technology, asking how students know whether they understand something and what to do when they don't understand.

Despite the scarcity of references to metacognition in the science literature, a number of common threads relevant to metacognition emerge when comparing the literature on science and that on reading. These parallels include discussions about the acquisition of process skills, as well as about constructivism and self-regulation, student errors and misconceptions, and textbook difficulty.

The Acquisition of Process Skills

Science educators have identified a core set of process skills that students taking science courses are expected to acquire (Carin & Sand, 1985; Carter & Simpson, 1978; Esler & Esler, 1985; Peterson et al., 1984). These skills include observing, classifying, comparing, measuring, describing, organizing information, predicting, inferring, formulating hypotheses, interpreting data, communicating, experimenting, and drawing conclusions.

Several authors have noted the close correspondence between these process skills and the skills of reading (Carin & Sand, 1985; Carter & Simpson, 1978; Esler & Esler, 1985; Peterson et al., 1984; Resnick, 1983). For example, in their methods text for teachers, Carin and Sand observe that discovery science and reading emphasize the same intellectual skills and thinking processes. They note, "You are currently engaged in teaching reading in your science program whether or not you realize it. When you help your students develop scientific processes you are also helping them develop reading processes" (p. 242). The teacher's manual for the Holt Science series (Abruscato et al., 1984) similarly stresses this link: "The teaching of reading and math skills and the teaching of science skills become one: the teaching of those critical thinking skills needed for all branches of learning" (p. xxi).

It should be clear that many of the science process skills also can be regarded as metacognitive skills. Consider the science process skill of formulating conclusions. Carter and Simpson (1978) argue that this skill corresponds to the reading skills of analyzing critically, evaluating information, recognizing main ideas and concepts, establishing relationships, and applying information to other situations. These skills certainly are similar to the standards of evaluation discussed earlier. Carter and Simpson also argue that the science skill of interpreting data corresponds to the reading skills of summarizing new information and varying the rate of reading to suit the text. Again, both of these skills may be regarded as metacognitive.

Some educators have criticized process-oriented science curricula for reducing the amount of reading required of students and thus possibly interfering with reading development (Bredderman, 1982). However, given the interconnections discussed above, emphasizing process skills may strengthen reading skills, including metacognitive skills. Indeed, comparisons of process-oriented curricula with traditional textbook curricula provide no evidence that students using textbook curricula are superior in reading achievement. In fact, they provide some evidence favoring process-oriented curricula (Bredderman; Shaw, 1983; Shymansky, Kyle, & Alport, 1982). Obviously, textbook use per se is not a major factor in improving reading skills in general or metacognitive skills in particular.

Constructivism and Self-Regulation

Constructivism is perhaps the most important theoretical parallel between the science and the reading literature to emerge in recent years. Psychologists and reading educators have demonstrated that readers actively construct meaning from text by using their prior knowledge to help them interpret incoming text information (Anderson, 1984). Science educators similarly have come to acknowledge the need for learners to assume an active role in acquiring new knowledge (Fisher & Lipson, 1986; Johnson, 1985; Osborne & Wittrock, 1983; Pope & Gilbert, 1983; Resnick, 1983; Finley and Champagne & Klopfer, this volume). As Pope and Gilbert state, "Knowledge is seen as being produced by transactions between a person and the environment. An emphasis is now placed upon the active person reaching out to make sense of events by engaging in the construction and interpretation of individual experiences" (p. 194).

Osborne and Wittrock (1983) describe a constructivist model adapted for science learning: "To comprehend what we are taught verbally, or what we read, or what we find out by watching a demonstration or doing an experiment, we must invent a model or explanation for it that organizes the information selected from the experience in a way that makes sense to us, that fits our logic or real world experiences, or both" (p. 493). The process of building such a model requires the metacognitive skills of evaluating the plausibility of the model and revising hypotheses if necessary.

A related parallel is the shared concern with promoting self-regulation. For example, Hawkins and Pea (1987) draw explicitly on the work of Vygotsky in advocating an instructional approach that fosters a transition from regulation by others to self-regulation. Johnson (1985) also discusses the importance of students' assuming control over their own learning of science: "When [students] learn that they have some control over what information they have access to, students can see themselves as responsible directors of their own learning, not as helpless receptacles for information others pour into them" (p. 34). Maehr (1983) criticizes much of science instruction for being too teacher-directed: "Insofar as directed learning diminishes the participation of the learner in the learning process, it will in the long run reduce achievement....Problems with directed learning are likely to be most severe in the areas of science education, where, above all else, the intent is to create an independent achiever" (p. 185).

An unusual discussion of self-regulation appeared in an article written for science teachers by Johnson (1985), who considers the theoretical contributions of Luria's brain research to science learning. The parallels between Luria's notions about executive control and current metacognitive theory are striking. According to Luria, the "executive system" develops with the maturation of the frontal lobes at about age 15. This system, says Luria, formulates programs of action and regulates performance and progress by comparing results with the original intention.

Teachers should be aware that Luria's claim that executive control does not develop until age 15 receives little support in the literature on the development of metacognitive skills. Students considerably younger than 15 years are capable of engaging in self-regulation. Teachers who read Johnson's discussion without taking these facts into consideration may mistakenly assume that attainment of self-regulatory skills is not a realistic aim for the elementary level student.

Student Errors and Misconceptions

Another parallel between the literature on science and that on reading is the recognition that error analysis can reveal a great deal about cognitive and metacognitive processes. Reading researchers have found that examining the kinds of errors people make while reading reveals important information about their underlying cognitive processing. Fisher and Lipson (1986) similarly argue that one goal of science instruction is to teach students to recognize and correct their own errors: "They are to acquire the skills of error management and be able to debug their own 'programs'" (p. 784). As noted earlier, recognition and correction of one's own errors are fundamental metacognitive skills.

Student misconceptions about science concepts have received considerable attention in the science literature. Students frequently have strong erroneous beliefs about specific concepts (Alvermann, Smith & Readence, 1985; Eaton, Anderson, & Smith, 1983; McDermott, 1982; Minstrell, 1982. See also Champagne & Klopfer, Finley, and Roth, this volume). The problem with such misconceptions is a metacognitive one. If students are unaware that they do not possess the correct relevant knowledge, they cannot clarify their understanding.

Student Problems with Textbook Comprehension

According to Collette and Chiapetta (1984), "Printed text materials remain the most widely used of all teaching aids in the science classroom." Heavy reliance on textbooks is apparent even at the elementary level (Teters, Gabel, & Geary, 1984). This emphasis is troubling to both science and reading educators, given that many science texts are difficult for students to comprehend (Collette & Chiapetta; Esler & Esler, 1985; Mayer, 1983; Thelen, 1984; Armbruster and Finley, this volume).

Textbook difficulty has many causes, including heavy vocabulary load, poor text structure, and exceptionally dense information. Hurd (1983) reported that some science textbooks used in the middle grades introduced as many as 2,500 technical terms and unfamiliar words, noting that this vocabulary load "is so great that it essentially precludes a conceptualization of scientific ideas and principles" (p. 2).

There is general agreement that the best way to deal with the problem of textbook difficulty is to provide advance classroom preparation. Methods texts stress that teachers should take the time to preteach vocabulary and activate prior knowledge in science lessons, as they do in reading lessons (Carin & Sand,

1985; Esler & Esler, 1985; Peterson et al., 1985; Thelen, 1984).

The problems of relying on textbooks are compounded by the fact that many students simply do not know how to read and study expository text effectively. As Gabel (1984) notes, "The reason why many students do not do well in science courses may be because science educators for too long have not taught students how to read science materials" (p. 585). Teachers often operate under the misguided assumption that students have the skills and background knowledge necessary to learn the material on their own.

Applications to Instructional Practice

Although most of the recommendations in this section are derived from the literature on metacognition and reading, the general principles apply to all domains of learning, including science.

Once we define metacognitive strategies as those that give students the skill to control their own cognitive activities, instructors should realize that they have been teaching metacognitive strategies without realizing it. Many of the science process skills routinely taught may be regarded as metacognitive strategies. Teachers must understand that they need not learn a whole new instructional approach to foster metacognitive strategies. They may need to gradually shift responsibility for implementing these strategies from themselves to the students.

Reciprocal Teaching

One effective instructional approach for enhancing children's metacognitive awareness and control is reciprocal teaching. Reciprocal teaching is an excellent way to foster independent use of strategies because each student is expected to serve as "teacher" in small group lessons. One such program, developed by Palincsar and Brown (1984), includes instruction in the use of four different strategies designed to monitor and improve comprehension of expository text: (1) predicting upcoming information, (2) asking questions about material that has just been read, (3) identifying and clarifying confusing information, and (4) summarizing as a means of self-review. The training program they developed was introduced to students over a period of several weeks, and clear gains in comprehension were obtained.

Palincsar and Brown's work on reciprocal teaching has already come to the attention of science educators. Fisher and Lipson (1986) recommend reciprocal teaching as an effective way of fostering self-correction strategies that will enhance comprehension of science material. Hawkins and Pea (1987) cite the potential of this approach for promoting self-regulation in science learning. Champagne and Klopfer, and Santa and Havens (this volume) each discuss specific uses of reciprocal teaching in the science classroom.

Explicitness in Strategy Instruction

Increasing children's knowledge about how, when, and why to regulate their own comprehension and learning is at least as important as increasing their regulatory skills. Students need to know how to use a particular strategy, why that strategy is useful, and when it should be used. Teachers must be explicit about the purposes of the strategies they would like their students to use. Psychologists have found that while children may be perfectly capable of using strategies taught to them, they are unlikely to do so spontaneously unless they realize the strategies' value. Paris, Cross, and Lipson (1984) have developed Informed Strategies for Learning, a program to enhance metacognitive awareness about reading.

The ability to reflect on our own cognitive processes is a crucial first step to becoming a strategic learner. Teachers can foster this skill by modeling for their students the thinking and problem solving approaches they use as they attempt a particular activity. It is generally agreed that such thinking aloud is an effective instructional approach for enhancing metacognitive skills (Bereiter & Bird, 1985; Palincsar & Brown, 1984). The think-aloud approach can be applied to enhance comprehension not only of science texts but also of demonstrations, hands-on activities, and experiments in the science classroom.

Questions for Evaluating Comprehension

One way to teach students how to evaluate their understanding is to approach the task with a focus on what makes text hard to understand. This approach helps students realize that comprehension involves an interaction between the reader and the text; it is not something that exists in the reader alone. Frequently, when students have trouble understanding a particular text, they are unwilling to acknowledge their difficulty for fear of appearing unintelligent. They often think the problem is entirely their fault when, in fact, it may be the text's fault or an interaction of text-based and reader-based factors. Text-based problems occur frequently, as analyses of basal and content area texts have shown (Anderson & Armbruster, 1984).

This reasoning can be applied in teaching elementary school students about the different kinds of things that can make texts hard to understand. For example, in demonstration lessons that I have done, I have introduced six questions for students to ask themselves as they read. Each question corresponds to one of the standards for evaluating comprehension discussed earlier. The questions, with the corresponding standards noted in parentheses are:

1. Are there any words I don't understand? (lexical)
2. Is there any information that doesn't agree with what I already know? (external consistency)
3. Are there any ideas that don't fit together because I can't tell who or what is being talked about? (propositional cohesiveness)
4. Are there any ideas that don't fit together because I can't tell how the ideas are related? (structural cohesiveness)
5. Are there any ideas that don't fit together because I think the ideas are contradictory? (internal consistency)
6. Is there any information missing or not clearly explained? (informational completeness)

Students can use these questions to guide their reading of expository text or to examine orally presented information. Given the inherent difficulty of many science concepts, students must have a means of assessing their understanding.

Recommendations from Science Education

In recognition of the goal of fostering independent learners, many science educators have agreed that students should be provided with the means for controlling their own learning activities. They have made various recommendations to help promote the use of effective strategies in the classroom, many of which are distinctly metacognitive (Dempster, 1984; Gwynn, 1987; Middleton, 1985; Norton & Janke, 1983).

Middleton (1985) compiled a list of study and test taking techniques that he recommends teachers distribute to their students. The list includes (1) skimming to decide which parts are hard and which are easy to understand, (2) planning how much study time is needed, (3) rereading if there is a failure to understand, (4) generating and answering questions about the text, and (5) identifying main ideas. Although Middleton's list provides a good starting point, it is important to remember that the teacher should model for the students how, when, and why to apply metacognitive strategies.

References

Abruscato, J., Fossaceca, J.W., Hassard, J., & Peck, D. (1984). *Holt science.* New York: Holt, Rinehart & Winston.

Alvermann, D.E., Smith, L.C., & Readence, J.E. (1985). Prior knowledge activation and the comprehension of compatible and incompatible text. *Reading Research Quarterly, 20,* 420-436.

Anderson, R.C. (1984). Role of the reader's schema in comprehension, learning, and memory. In R.C. Anderson, J. Osborn, & R.J. Tierney (Eds.), *Learning to read in American schools: Basal readers and content texts* (pp. 243-258). Hillsdale, NJ: Erlbaum.

Anderson, T.H., & Armbruster, B.B. (1984). Content area textbooks. In R.C. Anderson, J. Osborn, & R.J. Tierney (Eds.), *Learning to read in American schools: Basal readers and content texts* (pp. 193-226). Hillsdale, NJ: Erlbaum.

Andre, M.D.A., & Anderson, T.H. (1979). The development and evaluation of a self-questioning study technique. *Reading Research Quarterly, 14,* 605-623.

Baker, L. (1984a). Children's effective use of multiple standards for evaluating their comprehension. *Journal of Educational Psychology, 76,* 588-597.

Baker, L. (1984b). Spontaneous versus instructed use of multiple standards for evaluating comprehension: Effects of age, reading proficiency, and type of standard. *Journal of Experimental Child Psychology, 38,* 289-311.

Baker, L. (1985a). Differences in the standards used by college students for evaluating their comprehension of expository prose. *Reading Research Quarterly, 20,* 297-313.

Baker, L. (1985b). How do we know when we don't understand? Standards for evaluating text comprehension. In D.L. Forrest-Pressley, G.E. MacKinnon, & T.G. Waller (Eds.), *Metacognition, cognition, and human performance* (pp. 155-205). New York: Academic Press.

Baker, L. (1989). Developmental change in readers' responses to unknown words. *Journal of Reading Behavior, 21,* 241-260.

Baker, L., & Brown, A.L. (1984a). Cognitive monitoring in reading. In J. Flood (Ed.), *Understanding reading comprehension* (pp. 21-44). Newark, DE: International Reading Association.

Baker, L., & Brown, A.L. (1984b). Metacognitive skills and reading. In P.D. Pearson (Ed.), *Handbook of research in reading* (pp. 353-393). White Plains, NY: Longman.

Baker, L., & Zimlin, L. (1989). Instructional effects on children's use of two levels of standards for evaluating their comprehension. *Journal of Educational Psychology, 81,* 340-346.

Bereiter, C., & Bird, M. (1985). Use of thinking aloud in identification and teaching of reading comprehension strategies. *Cognition and Instruction, 2,* 131-156.

Bredderman, T. (1982). The effects of activity-based science in elementary schools. In M.B. Rowe (Ed.), *Education in the 80s: Science* (pp. 63-75). Washington, DC: National Education Association.

Brown, A.L., Armbruster, B., & Baker, L. (1986). The role of metacognition in reading and studying. In J. Orasanu (Ed.), *Reading comprehension: From research to practice* (pp. 49-75). Hillsdale, NJ: Erlbaum.

Brown, A.L., & Day, J.D. (1983). Macrorules for summarizing texts: The development of expertise. *Journal of Verbal Learning and Verbal Behavior, 22,* 1-14.

Brown, A.L., & Smiley, S.S. (1977). Rating the importance of structural units of prose passages: A problem of metacognitive development. *Child Development, 48,* 1-8.

Carin, A.A., & Sand, R.B. (1985). *Teaching modern science* (4th ed.). Columbus, OH: Merrill.

Carter, G.S., & Simpson, R.D. (1978). Science and reading: A basic duo. *The Science Teacher, 45,*(3), 20.

Collette, A.T., & Chiapetta, E.L. (1984). *Science instruction in the middle and secondary schools.* St. Louis, MO: Times Mirror/Mosby.

Dempster, L. (1984). An exchange of views on the place of reading in science instruction. *Journal of Reading, 27,* 583-584.

DiSessa, A.A. (1987). The third revolution in computers and education. *Journal of Research in Science Teaching, 24,* 353-367.

Durkin, D. (1984). Do basal manuals teach reading comprehension? In R.C. Anderson, J. Osborn, & R.J. Tierney (Eds.), *Learning to read in American schools: Basal readers and content texts* (pp. 29-38). Hillsdale, NJ: Erlbaum.

Eaton, J.F., Anderson, C.W., & Smith, E.L. (1983). When students don't know they don't know. *Science & Children, 20,* 6-9.

Elliott-Faust, D.J., & Pressley, M. (1986). How to teach comparison processing to increase children's short- and long-term listening comprehension monitoring. *Journal of Educational Psychology, 78,* 27-33.

Esler, W.K., & Esler, M.K. (1985). *Teaching elementary science* (4th ed.). Belmont, CA: Wadsworth.

Fisher, K.M., & Lipson, J.I. (1986). Twenty questions about student errors. *Journal of Research in Science Teaching, 23,* 783-803.

Flavell, J. (1981). Cognitive monitoring. In W.P. Dickson (Ed.), *Children's oral communication skills* (pp. 35-60). New York: Academic.

Forrest-Pressley, D.L., & Waller, T.G. (1984). *Metacognition, cognition, and reading.* New York: Springer-Verlag.

Gabel, D. (1984). An exchange of views on the place of reading in science instruction. *Journal of Reading, 27,* 585.

Garner, R. (1987). *Metacognition and reading comprehension.* Norwood, NJ: Ablex.

Garner, R. (1981). Monitoring of passage inconsistency among poor comprehenders: A preliminary test of the "piecemeal processing" explanation. *Journal of Educational Research, 74,* 159-162.

Garner, R. (1985). Text summarization deficiencies among older students: Awareness or production ability? *American Educational Research Journal, 22,* 549-560.

Glenberg, A., Wilkinson, A., & Epstein, W. (1987). Inexpert calibration of comprehension. *Memory & Cognition, 15,* 84-93.

Gwynn C. (1987). The well-read textbook. *The Science Teacher, 54*(3), 38-40.

Hawkins, J., & Pea, R.D. (1987). Tools for bridging the cultures of everyday and scientific thinking. *Journal of Research in Science Teaching, 24,* 291-307.

Hurd, P.D. (1983). Middle school/junior high science: Changing perspectives. In M.J. Padilla (Ed.), *Science and the early adolescent* (pp. 1-4). Washington, DC: National Science Teachers Association.

Johnson, V.R. (1985). Concentrating on the brain. *Science Teacher, 52*(3), 33-36.

Maehr, M.L. (1983). On doing well in science. Why Johnny no longer excels; why Sarah never did. In S. Paris, G. Olson, & H. Stevenson (Eds.), *Learning and motivation in the classroom* (pp. 179-210). Hillsdale, NJ: Erlbaum.

Mayer, R.E. (1983). What have we learned about increasing the meaningfulness of science prose? *Science Education, 67,* 223-237.

McDermott, L.C. (1982). Problems in understanding physics (kinematics) among beginning college students—with implications for high school courses. In M.B. Rowe (Ed.), *Education in the '80s: Science* (pp. 106-128). Washington, DC: National Education Association.

Middleton, J.L. (1985). I just can't take tests! *The Science Teacher, 52*(2), 34-35.

Miller, G. (1985). The effects of general and specific self-instruction training on children's comprehension monitoring performance during reading. *Reading Research Quarterly, 20,* 616-628.

Minstrell, J. (1982). Conceptual development research in the natural setting of a secondary school science classroom. In M.B. Rowe (Ed.), *Education in the '80s: Science* (pp. 129-143). Washington, DC: National Education Association.

Norton, D., & Janke, D. (1983). Improving science reading ability. *Science & Children, 20*(4), 5-8.

Osborne, R.J., & Wittrock, M.C. (1983). Learning science: A generative process. *Science Education, 67,* 489-508.

Otero, J. (1987, July). *Comprehension monitoring in learning from scientific text.* Paper presented at the Second International Seminar on Misconceptions and Educational Strategies in Science and Mathematics, Cornell University, Ithaca, NY.

Palincsar, A.S., & Brown, A.L. (1984). Reciprocal teaching of comprehension-fostering and comprehension-monitoring activities. *Cognition and Instruction, 1,* 117-175.

Paris, S.G., Cross, D.R., & Lipson, M.Y. (1984). Informed strategies for learning: A program to improve children's reading awareness and comprehension. *Journal of Educational Psychology, 76,* 1239-1252.

Paris, S.G., & Jacobs, J.E. (1984). The benefits of informed instruction for children's reading awareness and comprehension skills. *Child Development, 55,* 2083-2093.

Paris, S.G., & Myers, M. (1981). Comprehension monitoring, memory, and study strategies of good and poor readers. *Journal of Reading Behavior, 13,* 5-22.

Peterson, R., Bowyer, J., Butts, D., & Bybee, R. (1985). *Science and society: A source book for elementary and junior high school teachers.* Columbus, OH: Merrill.

Pope, M., & Gilbert, J. (1983). Personal experience and the construction of knowledge in science. *Science Education, 67,* 193-203.

Resnick. L.B. (1983). Toward a cognitive theory of instruction. In S. Paris, G. Olson, & H. Stevenson (Eds.), *Learning and motivation in the classroom* (pp. 5-38). Hillsdale, NJ: Erlbaum.

Shaw, T.J. (1983). The effect of a process-oriented science curriculum upon problem solving ability. *Science Education, 67,* 615-623.

Shuell, T.J. (1987). Cognitive psychology and conceptual change: Implications for teaching science. *Science Education, 71,* 239-250.

Shymansky, J.A., Kyle, W.C., & Alport, J.M. (1982). How effective were the hands-on science programs of yesterday? *Science & Children, 20*(3), 14-15.

Teters, P., Gabel, D., & Geary, P. (1984). Elementary teachers' perceptions on improving science education. *Science and Children, 22*(3), 41-43.

Thelen, J.N. (1984). *Improving reading in science* (2nd ed.). Newark, DE: International Reading Association.

Vygotsky, L.S. (1978). *Mind in society.* Cambridge, MA: Harvard University Press.

Watson, F. (1983). On the drawing board: A 21st century curriculum. *The Science Teacher, 50*(3), 62-63.

CHAPTER 2 *Science and Reading: Many Process Skills in Common?*

Michael J. Padilla, K. Denise Muth,
Rosemary K. Lund Padilla

This chapter clarifies the similarities between science process skills and reading comprehension skills and explains how they overlap. The authors describe the relationship between science process skills and problem solving in general, explain how scientific skills are integrated into scientific experimentation, and analyze reading as a problem solving activity. The chapter concludes with a visit to a classroom where a sample science lesson illustrates how reading and science process skills interrelate.

The following dialogue came from an introduction to a fourth grade science lesson in which the teacher sought to involve students in science process skills.

Teacher: Did you know that some parts of our skin are more sensitive than others? Joe, what part of your skin do you think is most sensitive?

Joe: I don't know.

Teacher: Can anyone help Joe out? (Silence.) Well, today we're going to make a sensitivity testing device. We'll call it our STD. We are going to use the STD on the skin of different parts of our body—knees, arms, backs of hands, and fingertips. Can anyone predict which part of our skin will be the most sensitive?

Mary: I'll bet the back of my knees are.

Teacher: Why?

Mary: I'm ticklish. (Laughter.)

Teacher: Okay, Mary, remind me to come back to your prediction after we finish using our STDs. Now, Raymone, tell me what we're going to do.

Raymone: I guess we're going to make these things and see what part of our skin feels the most.

Teacher: Feels the most?

Raymone: Yeah, like is my hand more sensitive than my arm?

Teacher: Great. I think you have the idea. Please come up and get a copy of the experiment. Materials are in the back of the room.

In this lesson, the teacher provided students with opportunities to apply reading process skills—which are essential for comprehending science texts—as well as science process skills. What does science process instruction have to do with teaching students how to read science texts? Are skills learned in science applicable to understanding the printed word? Are skills needed to comprehend expository text applicable to experimenting in science? This chapter explores these and other questions.

Science Process Skills and Problem Solving

In the *IDEAL Problem Solver,* Bransford and Stein (1984) write that a problem exists whenever the present situation is not the desired situation. Problem solving, then, is the act of trying to resolve problem situations so that answers can be found and progress made. The authors describe Bransford's IDEAL model for solving many types of problems. Each letter of the acronym stands for a separate step in problem solving. The first step is to identify the problem, which Bransford calls "the most significant part of problem solving" (p. 12). Next comes defining and representing the problem, important because "different definitions of the problem often lead to different treatments" (p. 15). The third step is exploring alternative approaches. The last two steps are acting on a plan and looking at the effects, which imply carrying out at least one of the potential solutions generated in step three and logically drawing conclusions from the results of this action.

Process skills, as defined and used in science education, mirror Bransford's model in two ways: (1) they are used to solve problems in science and (2) they follow a logical order, starting with problem identification and usually finishing with a set of conclusions. Several names have been used to describe science process skills: the scientific method, scientific thinking, and critical thinking skills.

Popularized by an elementary curriculum project called Science—A Process Approach (SAPA), science process skills are defined today as a set of broadly transferable abilities that are appropriate to many science disciplines and that reflect the true behavior of scientists. SAPA was built on the premise that process skills could be sequenced and taught in a hierarchical manner, beginning with the basic skills that serve as a foundation for the more complex, integrated skills.

Basic Science Process Skills

Observing, or using the senses to gather information about an object or event, is a basic process skill that young children use naturally. Later, they learn to classify objects and events by their distinguishing attributes. For example, children learn that all four-legged animals with tails and furry coats are not dogs. Classifying, like observing, makes direct use of the senses.

Inferring, on the other hand, draws on knowledge obtained through classifying and observing to make an educated guess about an object or event. For example, children may infer that dogs and cats are preferred over foxes as pets. They may even predict that a fox would be an inappropriate animal to have as a pet. When children communicate this acquired body of knowledge about pets to others through words or pictures, we say they are communicating—another basic science process skill.

Because observing, classifying, inferring, predicting, and communicating are essential to several of the steps outlined in the IDEAL problem solving model, science educators cannot assume that children will master these basic processes without instruction. Research suggests that students in grades 5 through 8 have not mastered some of the most basic science process skills (Padilla, Cronin, & Twiest, 1985). Students need many opportunities to apply these skills in meaningful instructional activities, whether in science or in reading classes.

Integrated Science Process Skills

Integrated skills are more complex than basic skills. Usually, integrated skills have to do with experimentation. Formulating hypotheses about the expected outcome of an experiment and identifying which variables will be held constant and which will be allowed to vary are two examples of integrated process skills. Other examples include defining how to measure a variable (e.g., stating that plant growth will be measured in centimeters per week), conducting an experiment, and interpreting the data in order to draw conclusions.

Integrated science process skills incorporate one or more of the basic skills and can overlap. For example, conducting an experiment from beginning to end involves observing, hypothesizing, identifying which variables to measure and how to measure them, interpreting the results, and communicating those results to others. Students who have not reached the formal operations stage of development reportedly have difficulty learning complex science concepts (Cantu & Herron, 1978; Padilla, Okey, & Dillashaw, 1983).

The results of studies involving middle school students have shown that integrated science processes can be taught (Padilla, Okey, & Garrard, 1984). However, for students to master these processes, they must have numerous opportunities to practice them in a variety of contexts and across different content areas. Also, teachers need to exercise patience in helping students learn how to incorporate the various basic and integrated process skills, especially given the developmental nature of the tasks involved.

Reading as a Problem Solving Activity

Whether conducting science experiments or reading assigned science texts, students are engaged in several of the same problem solving processes. The first step of experimenting in science involves identifying a problem and making observations about it. Similarly, reading begins with identifying the topic of a portion of text and then activating, or bringing to consciousness, relevant background knowledge about the topic. This problem identification step is considered the most important part of the IDEAL model of problem solving.

In science experimentation, as in reading, how skillfully the student negotiates the next step of the model—defining the problem—will affect the outcome of the experiment and the degree to which a text is comprehended. In science, how a student defines the problem will determine the variables selected for study and the treatments applied. In reading, the information a student predicts the author will present, and how well that prediction meshes with what actually is presented, will determine the rate and accuracy of comprehension.

Exploring alternative approaches, the third step in the IDEAL model, requires students to use their basic and integrated science process skills (classifying, predicting, formulating hypotheses) as well as their thematic reading processes (Kieras, 1985). Thematic reading processes (drawing conclusions, predicting outcomes) aid inferential comprehension and enable readers to distinguish between important and unimportant information. Locating the main idea and the supporting details are classification tasks; the reader identifies the main idea and organizes information that supports it. All these tasks are essential for exploring alternative approaches to solving a problem.

The integrated science process skills of carrying out an experiment, interpreting the results, and drawing conclusions are analogous to reading a text, making inferences, and drawing conclusions. Each plays a vital role in the final two steps of the IDEAL model: acting on a plan and looking at the effects. Acting on a plan as part of a science experiment means following through on at least one of the alternative approaches generated in step three. As part of reading a science textbook, acting on a plan means going forward with one of the predictions made in step three until it becomes obvious the prediction is incorrect and in need of revising.

Looking at the effects of a science experiment involves interpreting the results and drawing logical conclusions. Sometimes these conclusions help to formulate models or theories of how similar objects and events might perform in the future. At other times, such conclusions may point out the need to change the way certain variables are measured. Readers also are expected to interpret data (their textbooks) and draw logical conclusions from what they have read. Their conclusions will be influenced by their prior knowledge and by how closely the textbook information matches their previous concepts.

It would be naive to assume that a one-to-one relationship exists among all the science and reading processes described by researchers. Nevertheless, we believe that several critical similarities exist and that teachers should use these similarities to apply the skills taught in science to comprehension of written assignments. Likewise, we believe that treating reading as a problem solving activity provides students with opportunities to practice many of the science process skills they are expected to master.

Process Skills in Common

To bring our point home, we'd like you to complete a short exercise. Think back to the beginning of this chapter and the teacher who was attempting to involve her students in testing their sensitivity to touch. Figure 1 is a copy of the activity sheet the teacher used for her experiment. Read it and then list on a separate sheet of paper all the science and reading process skills this activity requires. Circle those skills that science and reading share. Then compare your list with ours as you read the rest of the chapter.

What reading and science process skills did this activity require? To successfully complete the activity, students had to make inferences, draw conclusions, make predictions, and verify their predictions. Examples of each of these processes follow.

Making Inferences

To assemble the sensitivity testing device, students had to infer several things about the directions. First, they had to infer that the straight pins were to be placed somewhere in the middle of the cardboard rectangle. Second, they had to infer that the pins needed to be placed lengthwise on the card in order to be 3 cm apart. Finally, they had to infer how to hold the STD once it was assembled. An important fact was not made explicit in the text of the lab sheet: feeling two pins on a particular body part probably meant that body part was more sensitive to touch than a body part where only one pin could be felt. This piece of information had to be inferred from the text.

Drawing Conclusions

Another process skill common to science and reading is drawing conclusions. In our example, students had to interpret their data before they could draw conclusions about the sensitive parts of their bodies. Specifically, they had to interpret the data recorded in their charts to determine which body part was most sensitive and why they came to that conclusion.

Making Predictions

Predicting was another common process skill used in the experiment. Students were asked to make two predictions. Before conducting the experiment, they were asked to predict which body part would be the most sensitive. To make reasonable predictions, they had to draw on prior knowledge about their bodies. After the experiment, students were asked to predict what would happen if they had no feeling in their arms. To do this, they had to draw on prior knowledge and then integrate that knowledge with both the informa-

Figure 1

Activity Sheet for Testing Sensitivity to Touch

Sensitive Points

Your skin is more sensitive in some spots than in others because some parts have more nerve endings. The nerve endings help us to feel. Let's see which part of our body is the most sensitive. Complete the following tasks.

1. Make a sensitivity testing device. Here's how. Take a piece of cardboard about 3 by 6 cm. Push two straight pins through the cardboard about 3 cm apart. You now have a sensitivity testing device, or an STD.
2. Test different parts of your body with the STD. Try your arm, your knee, the back of your hand, and your fingertips. To test a body part, blindfold yourself. Have a friend gently touch a body part with the STD. Do you feel one pin or two pins? Repeat five times for each part.
3. Collect your data on a table like this one. Record the times you felt the correct number of pins.

Body Part	Number of Times Correct

4. Before starting, predict which body part will be most sensitive. My prediction is______________________________.

   My reason for thinking this is ________________________

   ____

Making Sense

After finishing the test and filling in your table, answer the following questions.

1. Which body part was most sensitive? ________________ Did you correctly predict this? __________
2. Can you guess why this part was so sensitive? ________________

   ____
3. Predict what might happen if you had no feeling in your arm. _____

   ____

tion presented in the worksheet and the data they compiled from the experiment.

Verifying Predictions

Verifying predictions is a metacognitive skill shared by science and reading (Baker, this volume). This task is part of the knowledge and regulation of any cognitive task (Flavell, 1979). Students' metacognitive knowledge of a reading or scientific task involves being aware of whether they understand the material they are reading or the activity they are performing. Students' metacognitive regulation of the two tasks enables them to take corrective action if they do not understand the material they are reading or the science activity they are performing.

In the touch sensitivity exercise, students were asked to verify their predictions in the second part of question one under Making Sense. By verifying their predictions, students had an opportunity to correct any prior misconceptions they may have had about the sensitivity of body parts.

Conclusion

Making inferences, drawing conclusions, and making and verifying predictions are process skills that elementary and secondary teachers know how to teach. Elementary teachers have taught these skills in a myriad of ways, through basal reading selections, math lessons, and social studies projects. Secondary science teachers are content specialists. They know the value of process skills in teaching students how to think about the structure of a particular discipline, whether it be biology, physics, or geology. Equipped with this basic knowledge and an accompanying repertoire of strategies, teachers of science at both the elementary and secondary level need only to adapt the strategies they have used successfully in the past.

References

Bransford, J.E., & Stein, B.S. (1984). *The IDEAL problem solver.* New York: W.H. Freeman.

Cantu, L., & Herron, D. (1978). Concrete and formal Piagetian stages and science concept attainment. *Journal of Research in Science Teaching, 15,* 135-143.

Flavell, J.H. (1979). Metacognition and comprehension monitoring. *American Psychologist, 34,* 906-911.

Kieras, D.E. (1985). Thematic processes in the comprehension of technical prose. In B.K. Britton & J.B. Black (Eds.), *Understanding expository text* (pp. 89-107). Hillsdale, NJ: Erlbaum.

Padilla, M.J., Cronin, L., & Twiest, M. (1985). *The development and validation of a test of basic process skills.* Paper presented at the annual meeting of the National Association for Research in Science Teaching, French Lake, IN.

Padilla, M.J., Okey, J.R., & Dillashaw, F.G. (1983). The relationship between science process skills and formal thinking abilities. *Journal of Research in Science Teaching, 20,* 239-246.

Padilla, M.J., Okey, J.R., & Garrard, K. (1984). The effects of instruction on integrated science process skills achievement. *Journal of Research in Science Teaching, 21,* 277-287.

PART 2

Prior Knowledge and Science Text

The five chapters in Part Two center on two themes that challenge both science and reading educators: (1) the ways science texts are written make them difficult for students to understand, and (2) readers often do not possess adequate background knowledge for dealing with science concepts.

The questions addressed in this section relate to any elementary and secondary classroom in which reading plays a role in the science curriculum. Do science texts reflect the goals of science? Are they well written? How does the reader's prior knowledge influence learning? What happens when background knowledge is inadequate or when a reader's personal theory of why something happens in science is different from the way the concept is presented in the text?

After grappling with these issues, the authors provide teachers with workable solutions. In the end, the text, the student, and the teacher form an interlocking triangle. The teacher is at the apex of the triangle as the moderator in the interplay of student knowledge, text material, and learning strategies.

CHAPTER 3

Why Students Have Trouble Learning from Science Texts

Fred N. Finley

In this chapter Finley introduces the two issues that are at the heart of Part Two by describing problems with science texts and students' prior knowledge. He first explains why students have so much difficulty learning from science texts. Scientists want students to become keen observers and problem solvers, but most texts do not reflect this scientific model. Consequently, a clash of interests occurs that creates a difficult and often confusing learning situation for students. Finley also identifies several problems related to prior knowledge that make science teaching particularly challenging. He concludes by offering suggestions applicable to both elementary and secondary teachers.

Teachers at both the elementary and the secondary levels use a variety of methods to teach about the natural world. One of the most widely used methods is to assign students to read textbook material. Teachers assign this material on the assumption that the texts will provide students with much of the scientific information they need. They also assume that as students read, they will learn the ideas in the text and will be able to apply those ideas to a variety of tasks. Only the teachers' presentation of scientific knowledge is a more important source of information.

Although teachers may act as if the above assumptions are true, they know they are not. They complain that the texts are inadequate and worry about their students' limited ability to understand them. In short, teachers know that students have great difficulty learning from their science reading assignments.

Because teachers rely extensively on textbooks, it is important for them to understand why their students have such trouble with science texts. Once the possible explanations are considered, suggestions can be made for improving the writing and use of science texts.

The focus of this introductory chapter is broad by design. It sets the stage for a more detailed look at the instructional implications of inconsiderate or unfriendly texts (see Meyer and Holliday, this volume). It also provides a framework in which to discuss the instructional implications of students' lack of appropriate background knowledge—or, more specifically, their misconceptions (see Roth and Champagne & Klopfer, this volume).

The Nature of Science Textbooks

One of the reasons students have difficulty learning from science textbooks is that a fundamental difference exists between the nature of science textbooks and the educational goals those texts are meant to serve. Two major goals of science education are that students learn to describe and to explain natural phenomena. Part of the problem is that authors seldom provide well-written descriptions and explanations of phenomena. Given that Holliday (Chapter 5, this volume) expands on the problem of poorly written explanations, this chapter focuses strictly on the mismatch between the way concepts are explained and the goals of science education.

This mismatch between science education goals and the nature of text can be seen clearly in the following passage about molecules in motion:

> The kinetic theory is a useful model. If ammonia water is placed in an open dish in front of the room, you will smell the odor in a very short time. You will smell this odor even if there are no air currents in the room. How fast these molecules move depends upon the energy they have. The energy of motion of objects, including molecules, is called kinetic energy. The higher the temperature, the greater the speed of molecules, and so the greater their kinetic energy also.
>
> The statement that matter is made of atoms and molecules that are in constant zigzag motion is part of the kinetic theory of matter. This theory has been used to explain how matter behaves and to predict new facts about matter (Tracy, Tropp, & Freidl, 1983, pp. 52-53).

The stated objective of this passage is to explain the kinetic theory of matter. The authors want to make the key ideas scientists have developed understandable to the students. This objective results in a passage quite different from text that would be written with the objective of providing the scientific explanation for a particular set of phenomena (e.g., the behavior of gases). If the latter objective were adopted, the reader would see (1) a description of the set of phenomena to be explained, (2) the proposition that the phenomena are explainable in terms of moving molecules, (3) an actual explanation, and (4) information supporting the explanation. Furthermore, this material would be presented in a manner consistent with the structure of scientific explanations.

Instead, a single phenomenon—the smelling of ammonia—is presented. This phenomenon is incidental to the remainder of the passage; it is not presented as the phenomenon to be explained, and, in fact, never is explained. In addition, the text's claim that the concept of molecules in motion is part of the kinetic theory of matter is not adequately justified. For example, an important part of the justification—the applicability of the idea to explaining a variety of phenomena—is not presented.

This kind of text leaves students on their own to construct the actual explanations of phenomena such as smelling ammonia from across the room. Students also are on their own to develop an understanding of the structure of scientific explanations. Furthermore, because science texts typically fail to include the reasoning that supports scientific beliefs, students do not have the proper context to make sense of key ideas. Nor do they have reasons for believing the key ideas are valid and useful. Students are expected to set aside their own ideas and accept new ideas on the basis of the text's authority—a formidable task.

Problems Related to Prior Knowledge

Learning from text is likely to be successful when a student's knowledge of a topic is consistent with the information presented in the text, is fairly extensive, and is correct but not necessarily complete (Finley, 1990). Three factors involving prior knowledge can hamper learning from text. The most obvious is students' lack of knowledge about a topic. The second is students' incorrect prior knowledge, or misconceptions about a topic. The third has to do with the variability of students' knowledge.

Limited Prior Knowledge

If students don't know the meaning of terms and concepts contained in the text, they cannot construct an adequate understanding even if they can decode the words. Reading the following scientific material illustrates this situation:

> A cactolith is a quasihorizontal chronolith composed of anastomosing ductoliths, whose distal ends curl up like a harpolith, thin like a sphenolith, or bulge discordantly like an akomlith or ethmolith (McPhee, 1981, p. 27).

Most students probably can pronounce the words in this passage and recognize that the words should have meaning. However, few will understand the passage because they don't know what most of the words mean. The best readers can do is construct a partial interpretation of the selection based on any known words and on their understanding of the text's syntax.

The comprehension process probably functions as follows. The student decodes the words as she reads. She probably knows the meaning of quasihorizontal, curled up ends, thin, and bulge. This information, plus her knowledge of the text's syntax, may be sufficient to generate a partial understanding of a cactolith. That partial understanding may include the idea that a cactolith is quasihorizontal with curled-up ends that are either thin or bulge. Yet the information remains incomplete because the reader still would not know what a cactolith is. If she knew that the word ending lith meant rock, or body of rock, her understanding would be fuller but still largely inadequate.

The above example illustrates that readers construct their interpretations of science text around the words and concepts they know. Often they augment their interpretations by using their knowledge of the text's syntax or structure. However, this additional processing requires that a reader's initial understanding of the text be sufficient to allow for syntactic relationships to be identified and evaluated (Huggins & Adams, 1980).

A special case of failure to learn occurs when concepts are expressed in unfamiliar symbols or mathematical terms—particularly when students need to know these concepts in order to understand the rest of the text. If the symbols are not recognized, understanding is nearly impossible. Only those students with extensive prior knowledge of the subject will be able to construct the text's meaning by using context clues and syntax.

Science teachers at all levels must make sure their students have sufficient background knowledge to understand an assigned text selection. It is essential to take time before students read to find out what they already know about a topic and then explain the meaning of unfamiliar words and scientific symbols. The more difficult the concepts, the more extensive the instruction should be before students read.

Incorrect Prior Knowledge

The second circumstance in which students are likely to have difficulty learning from science text is when their prior knowledge is incorrect. Since Roth (Chapter 6, this volume) presents a thorough discussion of how students' misconceptions influence science learning, my comments here are brief.

Students seem to rely on their initial beliefs to construct an interpretation of text. They tend to maintain incorrect beliefs unless they recognize that the information in the text is inconsistent with their expectations. Unfortunately, research suggests that students rarely recognize the inconsistency between their ideas and those in the text. When readers have incorrect prior knowledge, their ideas override the textual information (Afflerbach, 1986;

Alvermann, Smith, & Readence, 1985; Lipson, 1982; Smith, Readence, & Alvermann, 1984).

Finley (1987) found that students whose initial knowledge was incorrect added from the text only information that could be construed as consistent with their ideas. For example, one student initially believed there was a substance that caused temperature changes but was unable to explain how the heat moved from one place to another. After reading a selection that explained temperature changes in terms of the transfer of molecular kinetic energy via molecular collisions, the student added the idea that heat is transferred from one molecule to another when molecules collide. He did so without giving up his belief in a heat substance. Finley's finding is especially important in light of recent research documenting that students often hold deeply rooted misconceptions about many natural phenomena (Ault, 1984; Champagne, Klopfer, & Anderson, 1980; Novick & Menis, 1987; Roth, this volume).

The Variability of Student Knowledge

If the content of students' prior knowledge is important, then a third feature must be considered—the extent to which knowledge varies within a class. This factor is essential because it relates to whether it is feasible to consider prior knowledge when trying to improve the writing and use of text. If each student's knowledge of a phenomenon were completely different, then writing or using text that accounts for that knowledge would be impossible. If all students' knowledge of phenomena were identical, then the task would be greatly simplified. As expected, the actual circumstances seem to be between these two extremes.

The results of one study illustrate the typical level of variability in students' knowledge of familiar phenomena. In a study by Finley (1985), high school chemistry students were asked to describe and explain an event involving the transfer of thermal energy by conduction. They came up with three possible explanations: one was correct but incomplete and two were fundamentally incorrect. Approximately one third of the ideas the students used were those needed to provide a complete and correct description of the event they were shown. Another third were idiosyncratic elaborations on one of the explanations or ideas they considered using but rejected quickly.

At first glance, this variability seems to confirm that writing and using text in ways that account for students' prior knowledge is impossible, but this is not the case. First, all students correctly described the transfer of thermal energy by conduction; second, one of the core explanations was correct and needed only to be completed; and third, only the two explanations that were fundamentally incorrect would need to be systematically addressed in the writing and use of text. If continuing research in other domains supports these findings, it will be clear that considering prior knowledge in the writing and use of text is feasible.

Conclusions

This chapter addresses several issues. First, textbook authors write in a way that is often inconsistent with the educational goals of science. Second, readers' prior knowledge influences their ability to recognize words and establish their meaning, to use syntax and text structure to relate previously unrelated ideas, and to construct an interpretation of the text content. If science educators wish to improve science textbooks and their use, they must consider students' prior knowledge and how it influences their explanations.

The next question is: What can be done? Science educators need to write a fundamentally different type of science text, one that provides rich descriptions and explanations of natural phenomena. Holliday (this volume) suggests ways to present science concepts more clearly. Yet writing with more clarity addresses only part of the problem.

Students also must be able to apply their background knowledge to the scientific explanations presented in texts. Where there is no match between students' prior knowledge and the information presented in textbooks, misconceptions occur. Texts need to be structured in ways that account for potential misconceptions. Meyer (this volume) presents a system that enables teachers to develop an awareness of how the texts they use are structured. In other chapters in this book, teachers will find helpful ideas for compensating for inconsiderate text and for creating conditions that promote conceptual change learning.

No matter how carefully texts are designed, teachers need to develop and employ instructional strategies that will assist students in learning from science texts. Several suggestions may be helpful. First, assess students' prior knowledge of the content. Have them write initial descriptions and explanations of phenomena; draw pictures and diagrams of sequences of events (such as the life cycle of plants) accompanied by written explanations of their drawings; present their ideas during class discussion so alternative descriptions and explanations can be considered; and interview small groups or individuals to elicit elaborated descriptions and explanations. When teachers have a good idea of what students know, they can take students' prior knowledge into account for instructional planning.

Once prior knowledge has been assessed, present phenomena to students in clear and direct ways. Use demonstrations, laboratory activities, field experiences, pictures, and films. Students need to know that they are to read for ideas that will help in explaining the phenomenon under investigation.

Third, complement the reading of text with instructional activities that provide students with evidence and reasons to question their initial beliefs. Use discrepant events to elicit discussion; then ask students to read their texts to develop more adequate descriptions and explanations.

Fourth, engage students in activities requiring them to apply their new knowledge in a way that will convince them that it is valid and useful for constructing the desired descriptions and explanations. Even when students understand new ideas, they may not be willing to set aside previous misconceptions. They need extensive experience.

In short, teachers can help students by developing assignments and leading discussions that make explicit students' initial interpretations of natural phenomena. Once students indicate what they know, teachers must challenge students' naive ideas by pointing out the information necessary for an explanation to be complete. Finally, after having students read their assignments, teachers can plan activities that will require students to apply their newfound knowledge in practical ways.

References

Afflerbach, P. (1986). The influence of prior knowledge on expert readers' importance assignment processes. In J.A. Niles & R.V. Lalik (Eds.), *Solving problems in literacy: Learners, teachers, and researchers* (pp. 30-40). Rochester, NY: National Reading Conference.

Alvermann, D.E., Smith, L.C. & Readence, J.E. (1985). Prior knowledge activation and the comprehension of compatible and incompatible text. *Reading Research Quarterly, 20,* 420-436.

Ault, C.R. (1984). The everyday perspective on exceeding unobvious meaning. *Journal of Geological Education, 32,* 89-91.

Champagne, A.B., Klopfer, L.E., & Anderson, J.H. (1980). Factors influencing the learning of classical mechanics. *American Journal of Physics, 48,* 1074-1079.

Finley, F.N. (1985). Variations in students' prior knowledge. *Science Education, 69,* 697-705.

Huggins, A.W.F., & Adams, M.J. (1980). Syntactic aspects of reading comprehension. In R.J. Spiro, B.C. Bruce & W.F. Brewer (Eds.), *Theoretical issues in reading comprehension* (pp. 87-112). Hillsdale, NJ: Erlbaum.

Lipson, M.Y. (1982). Learning new information from text: The role of prior knowledge and reading ability. *Journal of Reading Behavior, 14,* 243-261.

McPhee, J. (1981). *Basin and range.* New York: Farrar, Straus & Giroux.

Novick, S., & Menis, J. (1987). A study of student perceptions of the mole concept. *Journal of Chemical Education, 53,* 83-93.

Smith, L.D., Readence, J.E., & Alvermann, D. E. (1984). Effects of activating background knowledge on comprehension of expository text. In J.A. Niles & L.A. Harris (Eds.), *Thirty-third Yearbook of the National Reading Conference* (pp. 188-192). Rochester, NY: National Reading Conference.

Tracy, G.R., Tropp, H.E. & Freidl, A.E. (1983). *Modern physical science.* New York: Holt, Rinehart & Winston.

CHAPTER 4

Are Science Textbooks Considerate?

Linda A. Meyer

This chapter presents findings from Meyer's systematic analysis of science textbooks. The purpose of the analysis was to answer three questions: What information can be found in science textbooks? How is the content presented? How considerate is the text? Before learning the answers to these questions, you will be asked to engage in an exercise to help you distinguish the characteristics of considerate text from those of inconsiderate text. Although the chapter focuses on elementary science textbooks, both elementary and secondary teachers can use this analysis to select textbooks at their respective levels.

The procedure for analyzing science textbooks used in this chapter derives from the work of Armbruster and Anderson (1981), who have studied the characteristics of content area textbooks to determine what makes them considerate or inconsiderate. According to these authors, text is considerate when the writer has (1) systematically arranged the ideas in a pattern compatible with a particular discipline (e.g., cause/effect for science text); (2) logically connected the ideas; (3) avoided distracting or irrelevant information; and (4) taken into account the reader's probable background knowledge.

Inconsiderate text generally lacks one or more of these characteristics. It may assume the reader's background knowledge is more complete than it is or include irrelevant information. It also may include difficult technical terms, use unnecessary figurative language, or contain false information. Anderson and Armbruster count pictures and diagrams inconsiderate if they are unlabeled, unnecessary to the text, hard to see, or unclear in meaning.

Figures 1 and 2 present text segments on the topic of plant roots at the third grade level. The first segment was written to illustrate properties often found in inconsiderate text. The second segment—taken from a well written, widely selling basal science series (Mallinson et al., 1985)—is an example of considerate text. An exercise for comparing these two segments will demonstrate some of the characteristics that distinguish considerate text from inconsiderate text.

Directions for Completing the Exercise

After reading the text segments in Figures 1 and 2, divide a sheet of notebook paper in half lengthwise. Label the left column Text A and the right column Text B. In the left column, list all the inconsiderate features of Text A in red ink and all the considerate features in black ink. Do the same for Text B in the right column.

You also may want to describe the general characteristics of the two text segments, such as the number of sentences, photographs, and illustrations, where the questions appear (embedded in the text or separate from it), and whether the questions can be answered from the text or from prior knowledge.

After you have completed your analysis, decide which of the two text segments is more considerate. Then compare your findings with those described below.

Figure 1
Text A

Roots

The picture below shows two kinds of roots. How are they different? Could you pull one plant out of the ground more easily than the other?

Most plants have **roots** to hold them in the ground. If a plant does not have roots it can wash away or even blow away. Some roots have a big part. It may be fat or it may be long. Some roots have lots of parts that look like hairs. Those are skinny roots.

Roots collect water for plants to use. This means they **absorb** water. Rain goes into the ground. Some of the rain does not sink very far into the ground. Some of the rain sinks pretty far into the ground. Which of these roots would absorb water near the top of the ground? How do you know that?

Figure 2
Text B

Roots

What do roots do?

Have you ever tried to pull out weeds in a garden? Did they come out easily? Weeds and most other seed plants have roots that grow in the ground. In some plants, the roots may grow more than 6 meters deep. Roots hold plants in place. Look at these trees. Even a strong wind usually cannot blow them over.

Palm trees in a storm

What else do roots do? Roots take in the water and minerals (minər əls) plants need. Plants need minerals to grow and to be healthy. The water and minerals are carried in tiny tubes in the roots to the stem.

In some plants, food is stored in the roots. Have you ever eaten a carrot, radish, or beet? If so, you have eaten a root with stored food.

Different kinds of plants have different kinds of roots. Some plants have one large main root and other smaller roots. The large main root is called a **taproot.** Other plants have many roots that are all about the same size. These roots are **fibrous** (fībres) **roots.** Look at the drawing. Which plants have fibrous roots? Which plants have a taproot?

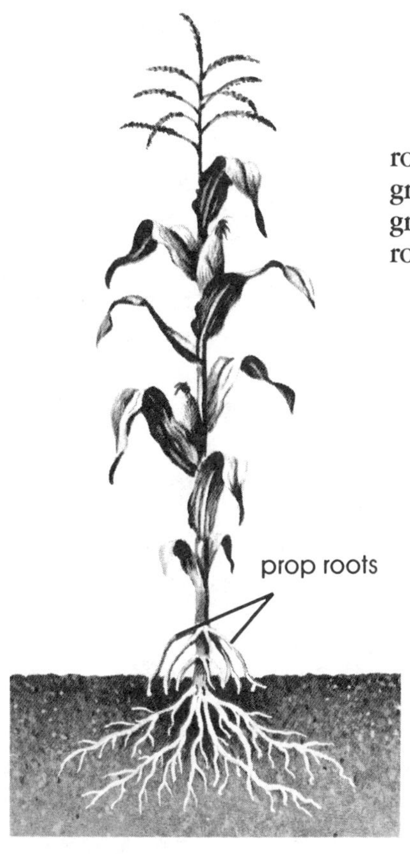

Another kind of root is the prop root. **Prop roots** are extra roots that grow out from the sides of stems. On some trees they grow downward from the tree branches. Corn plants and mangrove (mang grov) trees have prop roots. How do you think prop roots got that name?

Mangrove trees

Finding out

Does more of a plant grow in the ground, or above the ground? Get a potted plant. Take the plant out of the pot. Wash the soil off the roots with water. Then lay the plant on newspaper and carefully spread the roots and leaves.

Draw a circle around the roots. Draw another circle around the part of the plant that was above the ground. Cut along the lines you drew.

Compare the sizes of the two circles. Which circle is larger? Does more of this plant grow in the ground, or above the ground?

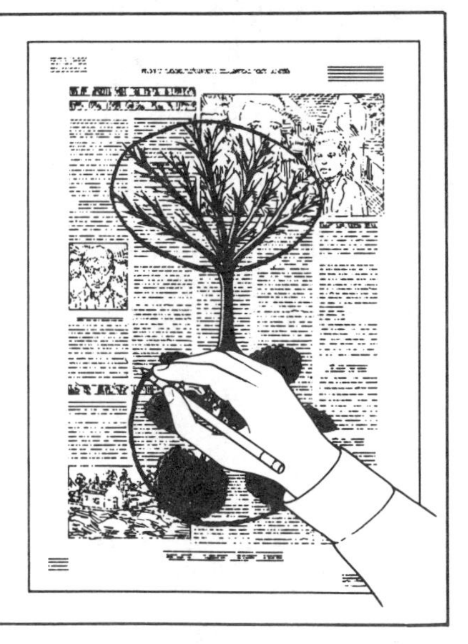

Results of the Analysis

Each figure shows the entire text segment on roots from its respective source. Text A (three paragraphs) has 16 sentences, 2 pictures, and 4 questions. None of the questions can be answered by reading the text, but two of them can be answered by referring to the pictures. The other two questions presumably can be answered from students' background knowledge. The passage contains no irrelevant statements.

Text A contains several examples of inconsiderateness. The first paragraph includes the broad question, "How are they different?" Students might come up with a variety of answers to this question and yet not be able to figure out which plant could be more easily pulled out of the ground (the next question in this paragraph). In the second paragraph, the idea that "if a plant does not have roots it can wash away or even blow away" is not logically connected to the next idea that "some roots have a big part." The lack of connectives at this point is somewhat jolting. In the third paragraph, a lack of connectives again results in text that does not flow easily from one sentence to the next. The pictures also are inconsiderate, as they are unlabeled.

In contrast, the segment on roots in Text B (five paragraphs and an activity) has 39 sentences, 8 pictures, and 11 questions. Five of the questions can be answered by reading the text, two can be answered by looking at the pictures, and four require students to come up with answers from their store of background knowledge. Text B also has no irrelevant statements.

The selection in Text B has more information and more opportunities for students to interact with the text because of the variety of places in which questions are asked. It also includes a meaningful activity in which students measure a plant's roots and compare them with the part of the plant that grows above ground. Students exposed to Text B learn not only what roots do but also that there are different kinds of roots. They see roots illustrated with pictures of common plants they may recognize from their background experience. All the photographs and illustrations are clearly labeled. In terms of writing, each paragraph begins with a statement or question that is then answered or explained in the sentences that make up the paragraph. In addition, Text B contains a greater repetition of vocabulary and makes better use of nouns and pronouns than Text A. The result is text that is better connected. In short, Text B is both more dense with information and better written.

A Larger Study

Because textbooks are used widely in science instruction (Finley, this volume; Mechling & Oliver, 1983), it makes sense to analyze their contents and characteristics carefully. Before the completion of our study (Meyer, Crummey, & Greer, 1988), no such systematic analysis existed.

We analyzed the textbook series in our study with a view toward answering three questions: What is the content? How is the content presented? How considerate is the text? In analyzing considerateness, we used the same procedure as that presented here for analyzing the two text segments on roots. In the overall analysis, however, we compared entire chapters of common content domains rather than a few pages on a specific topic.

The analysis was conducted on books chosen from four textbook series: the first through fifth grade science textbooks from two publishing firms (Textbook Series A and Textbook Series B), the fourth grade textbook from a third publisher (Textbook Series C), and the fifth grade textbook from a fourth publisher (Textbook Series D). Copyright dates for the four series range from 1974 to 1985.

These textbooks were selected for analysis for several reasons. They were in use in the three school districts participating in a longitudinal study of children's science concept acquisition and process learning, they represented programs implemented widely in elementary school classrooms in the United States, and they were representative of numerous other published elementary science textbooks. Although elementary texts were used in this analysis, the same procedure can be used to analyze secondary science texts.

Our goal was to explain why students in one district outperformed those in other districts on assessments measuring their knowledge of science concepts and processes. The first step toward reaching that goal involved determining the content and general properties of the four science textbook series under review.

Chapter or unit titles guided the first pass through the four textbook series to identify each content domain. An exception to this procedure was needed for the series in which only the fifth grade book was used. With that book, it was necessary to read each page to determine the content because the chapter and unit headings often were unclear or misleading. The vocabulary count in the domain analysis consisted of words the authors specified the students were to learn. Most of these words were listed, defined, and highlighted in each textbook.

In analyzing how the content was presented, we focused on content domains common across grade levels and publishers. We used this procedure instead of analyzing every page of each textbook because we believed that like-content would yield more defensible differences among series than would comparisons across total series. We established lecture/discussion and hands-on activities as the primary activities to tally from the teachers' editions. Optional activities and hands-on activities were tallied in the students' materials.

A further analysis of how the content was presented consisted of coding each question from the teachers' guides using Pearson and Johnson's (1978) categories of background knowledge, text-implicit, and text-explicit questions. In addition, we added a review question category as a measure of each series' attempt to reinforce previously taught ideas. We used the questions in the teachers' guide as measures of how the content was presented because we believe that questions establish the framework in which teachers interact with students after the lesson's content has been presented. Therefore, a textbook series dominated by questions students must answer from their background knowledge places heavy emphasis on the information students bring to each lesson. A series dominated by text-implicit and text-explicit questions, on the other hand, places greater emphasis on the text.

Results of the Larger Study

What is the content? We found major differences among the textbook series based on the number of content domains represented and the number of vocabulary words targeted for instruction, especially at the early elementary level. For example, at the first and second grade levels, Textbook Series B had 8 and 10 content domains, with 241 and 277 vocabulary words, respectively. Textbook Series A, on the other hand, had 4 and 9 content domains at those same levels, with only 56 and 44 vocabulary words.

A shift in the number of content domains occurred between these two publishers at the third and fourth grade levels. Series B had 15 content domains at both grade levels, whereas Series A jumped to 20 content domains at third grade and 25 content domains at fourth grade. In contrast, Series C had 19 content domains at the fourth grade level.

The vocabulary patterns found in Series A and B at the first and second grade levels held for the third and fourth grade levels. For these latter grades, Series B had 243 and 288 vocabulary words, respectively, while Series A presented 96 words at the third grade level and 137 words at the fourth grade level. Series C had 224 vocabulary words at the fourth grade level.

At the fifth grade level, Series B again had the lowest number of content domains (15) and the highest number of vocabulary words

(466) of the four series analyzed. Series A had 25 content domains and 210 vocabulary words at this level, while Series D had 24 content domains and 211 vocabulary words.

An analysis of content domains for these four publishers showed only three common content domains for grades 1, 2, and 3, and only four common content domains for grades 4 and 5. Thus, the content presented in these science textbook series is more diverse than similar.

How is the content presented? At the first, second, and third grade levels, Textbook Series A had far more lecture/discussion activities listed in the teachers' editions than did Textbook Series B. The ratio was approximately 5 to 1, with about 200 more lecture/discussion activities per grade level in Series A than in Series B.

At these same grade levels, however, Textbook Series B had more hands-on activities for teachers to direct. The ratios of these activities from Series A to Series B ranged between 3 to 1 and 10 to 1.

At fourth and fifth grade levels, the pattern established in the earlier grades persisted, although the differences were not as dramatic. Series B continued to have fewer lecture/discussion activities and more hands-on activities than did Series A.

We differentiated between optional activities and hands-on activities in the students' materials because those activities identified as optional were less likely to get done than those specified as hands-on activities. In first through fifth grade textbooks, Series A had the greatest number of optional activities, usually 5 or 6 times the number in Series B.

Table 1 shows the number of background knowledge, text-explicit, text-implicit, and review questions found in the four textbook series. Series B had a greater number of background knowledge questions than Series A in grades one, two, and three. At grades four and five, the order was reversed. In these grades, Series B had fewer questions that required background knowledge than did Series A, C, or D. Of the four textbook series analyzed, Series B had the greatest number of text-explicit and review questions at all five grade levels.

How considerate is the text? Overall, the four textbook series were quite considerate, as evidenced by the information contained in Table 2. For example, Series B averaged fewer than six instances of missing connectives or unclear referents per grade level. Only two instances of illogical structure were identified in all the chapters of this series. Series B also had only one example of an illogical sequence, explanation, or procedure.

It is important to keep in mind that the results in Table 2 are based on an analysis of 57 chapters, for which every word and picture were considered. Common content domains for these series were plants and animals or plants in grades 1 through 3; electricity and magnetism, the human body, and the solar system in grade 4; and the human body and weather in grade 5.

The only categories consistently troublesome across textbook series were incomplete background knowledge and problems with pictures and diagrams. A large number of pictures and diagrams were judged to be unnecessary, hard to see, or unclear.

Particularly important when looking across publishers in these common content domains is the number of zeros for each category. Every line in the table represents a grade level of chapters. There are 46 zero entries for the grade levels. Therefore, we found no examples of inconsiderateness in over 38 percent of the categories.

Implications for Practitioners

The contents of the four science textbook series we analyzed varied substantially. Series B had considerably more content than did Series A. Instructional materials that contain large amounts of content force teachers to work hard to cover that content or to make choices

Table 1
Types of Questions in Four Science Textbook Series*

Textbook Series	Background Knowledge Questions	Text-Explicit Questions	Text-Implicit Questions	Review Questions
		Level 1 (First Grade)		
B	315	141	183	205
A	127	91	157	22
		Level 2 (Second Grade)		
B	253	308	263	164
A	46	70	354	81
		Level 3 (Third Grade)		
B	127	552	183	321
A	97	50	493	96
		Level 4 (Fourth Grade)		
B	313	611	187	337
A	368	303	238	219
C	335	363	355	178
		Level 5 (Fifth Grade)		
B	208	712	232	368
A	234	179	296	253
D	618	482	139	247

*The four textbook series (in alphabetical order) are Abruscato et al. (1980), Holmes, Leake, & Shaw (1974), Mallinson et al. (1985), and Sund, Adams, & Hackett (1982).

about which activities or content domains to skip. By recognizing that science textbook series differ substantially in the amount of content they present, teachers, reading specialists, administrators, and curriculum directors will be alerted to the need to make adjustments in their pacing and assessment of instruction.

It is surprising that there is so little overlap in content among the four textbook series. Plants, animals, and living things are the only common content domains across two of the textbook series at the first and second grade levels. At the third grade level, the only common content domain is plants. The absence of

Table 2
Cross-Publisher Comparisons of Inconsiderate Text Structure and Content

Grade Level	Common Content Domains	Chapters	Number of Structure Problems			Number of Content Problems					Number of Problems with Pictures & Diagrams	
			Illogical Structure	Lack of Connectives or Unclear Referents	Illogical Sequences, Explanations, or Procedures	Irrelevant Ideas	Incomplete Background Knowledge	Problematic Technical Terms	Unnecessary Figurative Language	False Information	Unnecessary	Hard to See or Unclear
Textbook Series A (1982)												
1	Plants; animals	4	5	9	0	6	8	0	0	0	2	35
2	Plants; animals	2	2	11	0	1	20	0	0	0	5	35
3	Plants	2	4	9	0	6	15	0	0	0	9	10
4	Electricity and magnetism; human body; solar system	7	2	8	1	2	30	3	2	0	9	55
5	Human body; weather	6	1	3	0	10	28	7	1	0	9	4
	Mean (standard deviation)	4.2(2.3)	2.8(1.6)	8(3)	0.2(0.4)	5(3.6)	20.2(9.1)	2(3.1)	0.6(0.9)	0	6.8(3.2)	27.8(20.8)
Textbook Series B (1985)												
1	Plants; animals	4	0	0	0	2	22	2	0	0	0	19
2	Plants; animals	2	0	6	0	7	36	2	0	0	0	23
3	Plants	2	0	1	1	1	10	0	0	0	0	0
4	Electricity and magnetism; human body; solar system	7	2	10	0	6	25	4	0	0	10	21
5	Human body; weather	6	0	12	0	5	9	0	0	0	11	16
	Mean (standard deviation)	4.2(2.3)	0.4(0.9)	5.8(5.3)	0.2(0.4)	4.2(2.6)	20.4(11.2)	1.6(1.7)	0	0	4.2(5.8)	15.8(9.2)
Textbook Series C (1982)												
4	Electricity and magnetism; human body; solar system	13	0	6	0	13	33	6	1	0	21	47
Textbook Series D (1974)												
5	Human body; weather	2	6	13	0	8	15	0	1	0	8	10

a common core of content domains must be taken into account by curriculum directors and associate superintendents for instruction. Guidance counselors and educational psychologists also must consider the impact of this lack of uniformity on students' performance on standardized achievement tests.

Although one might expect a textbook series to emphasize either lecture/discussion activities or hands-on activities, no such dichotomy appeared in our analysis. This does not mean that such a split may not be present in other science textbook series. Individuals who have the responsibility of selecting science textbooks for their districts should ask themselves the following questions:

1. Are hands-on activities more important or more in line with the district's philosophy and goals than lecture/discussion activities?
2. Or are the two types of activities of equal importance?
3. Do they need to be integrally related?

Once teachers know which activity types they want to stress, they can determine the characteristics of several textbook series by sampling common content domains.

On the basis of this study's results, it appears that elementary science textbooks are not as inconsiderate as we may have been led to believe. Furthermore, the accuracy of information found in these textbooks is reassuring. This considerateness may encourage practitioners to use textbooks more and to assign more passages to be read independently. That students can be expected to learn difficult science concepts on their own, with little or no teacher direction, is not a reasonable implication to draw from the present analysis, however. Given the relatively high number of problems with incomplete background knowledge and unclear pictures and diagrams, teachers may wish to focus on building background knowledge in new content domains before students read their science assignments. Teachers also might plan ways of drawing students' attention to the information they can learn by integrating what they see in pictures with what they read in the body of the text.

Conclusion

The primary message of this chapter is that analyzing science textbooks in a systematic way yielded data quite different from what we had anticipated. This is an important point because it illustrates the need to gather careful descriptive data that can refute, verify, or modify some of our intuitions about textbooks. The results of this study also point out the danger of flipping through textbooks too quickly before coming to a conclusion. Problems may exist in these books, but the number of problems seems relatively small.

Practitioners can gather reliable information about science textbooks in a short period of time by selecting a content domain common to several textbooks at the same grade level and then asking: What is in the books? How is the content presented? How considerate is the text?

References

Abruscato, J., Fossaceca, J.W., Hassard, J., & Peck, D. (1980). *Holt elementary science.* New York: Holt, Rinehart & Winston.

Armbruster, B.B., & Anderson, T.H. (1981). *Content area textbooks* (Reading Education Report No. 23). Champaign, IL: University of Illinois, Center for the Study of Reading.

Holmes, N.J., Leake, J.B., & Shaw, M.W. (1984). *Gateways to science.* New York: McGraw-Hill.

Mallinson, G.G., Mallinson, J.B., Smallwood, W.L., & Valentino, C. (1985). *Silver Burdett science.* Morristown, NJ: Silver Burdett & Ginn.

Mechling, K.R., & Oliver, D.L. (1983). Activities, not textbooks: What research says about science programs. *Principal, 62*(4), 41-43.

Meyer, L.A., Crummey, L., & Greer, E.A. (1988). Elementary science textbooks: Their contents, text characteristics, and comprehensibility. *Journal of Research in Science Teaching, 25*(6), 435-463.

Pearson, P.D., & Johnson, D.D. (1978). *Teaching reading comprehension.* New York: Holt, Rinehart & Winston.

Sund, R.B., Adams, D.K., & Hackett, J.K. (1982). *Accent on science.* Columbus, OH: Charles E. Merrill.

CHAPTER 5

Helping Students Learn Effectively from Science Text

William G. Holliday

Initially readers may find that Holliday's chapter on the "unfriendliness" of text stands in stark contrast to Meyer's conclusions. But a second look reveals that his findings differ because his focus differs. Holliday critiques content dimensions, noting that most published questions do not require students to think much and emphasizing the need for clearer explanations. He wants authors to include better visuals and to delete extraneous information. Science texts are becoming more cumbersome; they contain too much information and include unnecessary scientific jargon. Holliday offers both elementary and secondary teachers practical advice on ways to deal with these problems.

In the previous chapter, Meyer portrays an optimistic view of elementary science texts. She found that structually, these texts appear relatively considerate. However, several nonstructural aspects of texts remain troublesome. Namely, texts are overloaded with verbatim recall questions, inadequate explanations, and irrelevant scientific jargon.

These problem areas have a direct bearing on how well students learn science content, particularly if science learning is to reflect the goals of scientific inquiry (Finley, this volume). If teachers want their students to develop rich descriptions and rational explanations of natural phenomena, they must ask more thinking questions, provide adequate explanations, and eliminate scientific jargon.

Using Thinking Questions

Poorly developed end-of-chapter questions and test items accompanying science texts usually require little thought (Prawat, 1989). Typically students can supply answers by looking up key words or by simply recalling and rearranging groups of words, numbers, or scientific symbols (Holliday & Benson, 1991). Such tasks and procedures often are meaningless. Questions that require little thought encourage students to perceive school science as a boring memory course (Holliday & McGuire, in press).

Of course, some memory-level learning is necessary for learning science. Factual recall exercises can help students identify information and learn facts they can apply to new situations (Shuell, 1990). But such factual questions should be used sparingly.

Nonthinking Text Questions

Both elementary and secondary school science texts contain numerous examples of nonthinking questions. For instance, the sentences displayed in Figure 1 come directly from a best selling science series.

Compare these sentences to the review exercises in Figure 2. Notice the lack of substantive differences between the prose and exercise wording. How much thinking is required of fifth graders to answer these matching items about electricity?

Figure 1
Sentence Definitions from a Fifth Grade Text

The path on which the electrons move is called a circuit.

A generator changes energy of motion into electrical energy.

A parallel circuit is one in which electrons can follow more than one path.

An atom with no charge is neutral.

Small amounts of electric power are measured in units called watts.

A series circuit is one in which the electrons can follow only one path.

An electric cell is a device that changes chemical energy to electrical energy.

Matter through which electrons move easily is called a conductor.

Silver Burdett Elementary Science, Level 5, pp. 154-177, 1987.

Figure 2
End-of-Chapter Exercises from a Fifth Grade Text

Write the letter of the term that best matches the definition. Not all of the terms will be used.

1. Unit for measuring small amounts of electric power
2. Circuit in which electrons can follow only one path
3. Matter through which electrons move easily
4. Device that changes chemical energy to electrical energy
5. Atom that has no charge
6. Path through which electrons move
7. Circuit in which electrons can follow more than one path
8. Machine that changes energy of motion into electrical energy

a. circuit
b. generator
c. parallel circuit
d. static electricity
e. neutral
f. watt
g. current electricity
h. series circuit
i. electric cell
j. conductor
k. turbine
l. insulator

Silver Burdett Elementary Science, Level 5, p. 178, 1987.

Elementary school students engaged in such verbatim-matching tasks can answer these questions by flipping pages. Rather than thinking about these concepts, students have only to define words presented in the prose with the same key words used in the exercises. Such search-and-compare tasks discourage real thought (Holliday, Wittaker, & Loose, 1984).

The problem is not confined to elementary science texts. A quick review of junior high and high school texts reveals similar types of review exercises. For example, ninth graders answering questions in one popular physical science text are instructed to classify chemical compounds as either acids or bases. Students can do this task readily by remembering the simple rule that chemical formulas containing the letter OH represent bases and all other formulas represent acids. They do not have to understand the important characteristics of acids and bases; they need only to apply a nonthinking rule.

Moreover, high school students using a popular biology text seldom think when answering the practice review questions. For example, most chapters conclude with a review page containing about two dozen vocabulary words, one dozen verbatim-style questions, and one or two thinking-style questions. Teachers often recommend that students know these terms and answer selected questions. To students, "knowing these terms" usually means copying word-for-word definitions directly from the text. Such exercises are not only a waste of time but also are painfully boring and often counterproductive. Copying chunks of words from text onto lined paper fails to facilitate increased understanding of important ideas and misleads students into believing they really know the terms (Holliday et al., 1984).

Troublesome Test Items

Test items contained in booklets accompanying texts can further reinforce students' misconceptions about the nature of school science and school learning. Not only do many items contain identical chunks of text information, but some even encourage the learning of irrelevant information that may foster misunderstanding.

For example, a test booklet accompanying one popular high school text contains many verbatim-recall items requiring only the memorization of a word or drawing unique to this biology text, with no other application. In one case, the authors used *protrusion* to define the term *alveoli* ("a protrusion in the air sac of the lungs where gases are exchanged").

Students who merely memorized *protrusion* in association with *alveoli* were fully prepared to answer this matching test item, and ultimately were reinforced. But they got the answer right for the wrong reason. Because of such school experiences, students learn early that memorizing the exact words in the text constitutes an adequate preparation for many test items (Andre, 1979).

Another example in the same booklet was a visual test item that presented an unlabeled drawing of the human heart. This drawing was identical to the one used in the text. It illustrated the heart's essential components, including the four chambers, associated valves, and external blood vessels. It also contained irrelevant visual characteristics, including an orientation that accented views of some portions of the heart, as well as graphic techniques that emphasized the border between adjacent structures. Such test items encourage students to learn text drawings by memorizing any visual characteristics, relevant or irrelevant. They learn to use whatever characteristics make their learning task easy and successful in terms of answering visual test items.

In this regard, research supports a well known maxim: students seldom do more work than the teacher requires (Andre, 1979). Psychologists often refer to this phenomenon as the least-effort principle (a principle that applies not just to students but to most of us).

Demanding Better Questions

The good news is that publishers are beginning to improve the quality of questions. For example, the verbal item about alveoli and the visual item about the heart are absent from the new edition of the text. However, the problem of poor text and test questions is far from solved.

Textbook adoption committees and publishers need to form closer working partnerships. Publishers might improve the quality of their questions if adoption committees encouraged such action. Teachers must apply political and instructional pressure to members of text adoption committees in their states, as well as representatives of publishing houses at conventions for science instructors. Teachers must insist that publishers upgrade the quality of questions contained in texts and in test booklets.

Study and test questions could become powerful instructional and evaluative tools if they were designed to encourage and evaluate thinking. Again, teachers must work with publishers. When teachers become more vocal, publishers will listen. They want to create best selling products, and excellence sells.

Teaching Recommendations

In the interim, teachers can use several classroom strategies to offset text problems. First, they can inform students that only thinking-style questions will appear on science tests. Many of these questions should focus on describing and explaining natural phenomena. Students need to understand that answering verbatim-style exercises or merely memorizing words and visual clues will not adequately prepare them for tests.

Second, teachers can show students how to answer thinking questions by producing and distributing sample tests and providing instruction on the best way to study for them. This instruction should include practice review questions containing different words, formats, and contexts from the ones presented in textbook explanations.

Students can inspect examples of thinking and nonthinking exercises for discriminating characteristics. They also can convert traditional exercises to thinking questions and explain why their new questions measure real thinking. Better yet, students can create their own questions focusing on key science concepts. Teachers can guide students in developing questions that require explanations of important scientific phenomena, and encourage them by including some of their questions on tests. Students can use their questions to monitor their own comprehension (Baker, this volume). Once students understand how to generate their own scientific questions, they become more responsible for their own learning.

Providing Adequate Explanations

The most common complaint about science texts is their lack of adequate explanations of important content (Armbruster, 1984; Finley, this volume). Poorly developed explanations cause confusion and can lead to incorrect ideas about important science concepts (Roth, this volume).

Herman (1984) examined the effects of inadequate explanations on students' learning. She presented eighth grade students with two versions of an explanation about how the human heart works. The original version, published in a popular junior high text, was tersely written. The revised version was designed to provide students with a more detailed explanation of the same topic. It contained more detail and an explanation of how various parts of the heart relate. Text passages 1 and 2 show the original and revised explanations.

As predicted, students learned more about the heart from the revised version. Perhaps the authors or publishers of the original version wrongly assumed that students could visualize an operating heart. Students had to guess or infer information such as the internal and external shape and size of the cardiac muscle, the logical flow of blood within the muscle, and the precise function of each chamber. The experimental findings showed that these eighth graders benefited from a more explicit, comprehensive explanation with fewer implicitly vague sentences. They needed a detailed description of how the chambers fill with blood to understand how the heart pumps blood from one compartment to another or to external blood vessels. They also needed an explanation of how the chambers are oriented in relation to one another and to external vessels.

In addition, the results of the study supported the need to clarify content by deleting extraneous information. For instance, consider how little the following analogy contributes to understanding: "The work done by the heart each minute is about equal to lifting 32 kg a distance of 30 cm off the ground." Such statements fail to help students while devouring valuable space that could be used for clear explanations and illustrations of important concepts.

The numerous instances of poorly developed explanations in science texts illustrate the need for science teachers to develop their own explanations of important concepts. For example, imagine yourself as a fifth grader reading this passage about the structure and function of a dry cell:

TEXT PASSAGE 1

Original Version

A human heart is a cone-shaped, muscular organ about the size of a large fist. The heart is located in the center of the chest behind the breastbone and between the lungs.

A human heart containes four chambers—right atrium (AY tree uhm), left atrium, right ventricle (VEN trih kuhl), and left ventricle. Right and left refer to the body's right and left sides. A wall separates the chambers on the right from the chambers on the left (Heimler & Lockard, 1977).

TEXT PASSAGE 2

Revised Version

The heart is the part of the circulatory system that pumps blood throughout the body. The heart is located in the center of the chest behind the breastbone and between the lungs. The human heart is suited for pumping because it is a hollow, cone-shaped, muscular organ about the size of a large fist. Being hollow, the heart can easily fill up with blood. Once filled, the heart muscle provides the power necessary for pumping the blood through the body.

A human heart contains four hollow chambers made for receiving and sending blood. The right atrium (AY tree uhm) and right ventricle (VEN truh kuhl) receive and send blood to the lungs, while the left atrium and left ventricle receive and send blood to the rest of the body. (Note that right and left refer to your body's right-hand and left-hand sides.) The right and left sides of the heart are separated by a wall of muscle. This wall keeps blood going to the lungs separate from the blood going to the body (Herman, 1984).

TEXT PASSAGE 3

One type of electric cell is called a dry cell. A dry cell uses a chemical paste, carbon rod, and zinc case to produce a flow of electrons. Chemical reactions occur inside the dry cell. One reaction causes the walls of the zinc case to become negatively charged.

Another reaction causes the carbon rod to become positively charged. The zinc case is called the negative pole. The carbon rod is called the positive pole. If the dry cell is connected to a circuit, electrons flow from the negative pole to the positive. From this movement of electrons flows electric current (Silver Burdett Elementary Science, Level 5, 1987).

A number of mystifying questions must come to the minds of fifth graders reading this passage. Some possible questions are: How can there be a chemical reaction without moving material? How can a chemical reaction, which ordinarily produces other chemicals and colors, produce electricity? How can electrons go from the outside of the battery (negative pole) through a wire and back into the center of the battery (positive pole) without depleting the electrons on the outside and overflowing the inside? What do the words *dry, paste, rod, case, flow, charged, pole, circuit,* and *movement* really mean in this context? It is a wonder that any student can understand, much less answer, thinking questions related to this explanation. Other popular texts designed for fifth graders contain equally inadequate explanations.

Better Visual Examples

One way to improve text explanations is through improved visuals (Holliday, 1990). Junior high students learning about glaciers, for instance, would benefit from both prose descriptions and visual examples, such as a series of illustrations showing how glaciers form and reshape the landscape. For example, in one earth science text a three-part drawing succeeds in illustrating the effects of glaciation (Figure 3).

Many texts do not have adequate visuals to explain concepts. For instance, in one earth science text, the concept of water erosion is

Figure 3

Glacial erosion of mountain peaks often leaves steep, jagged formations.

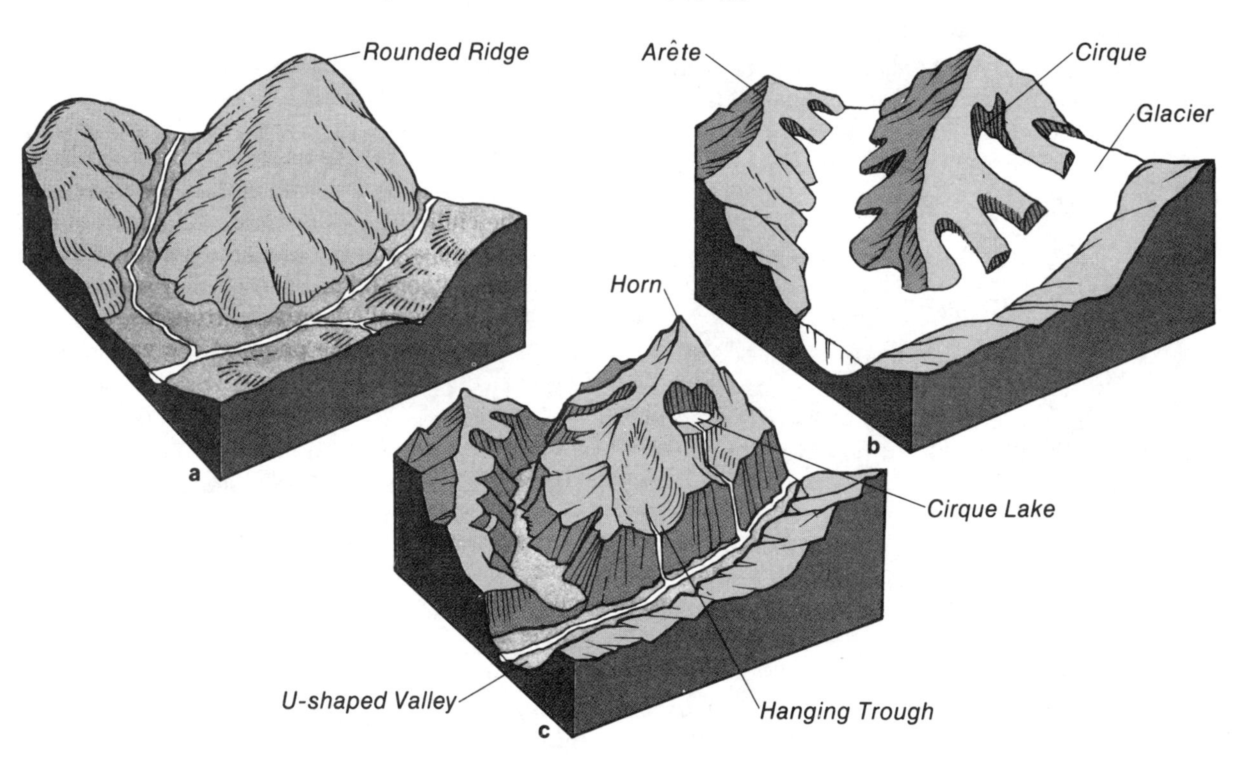

S. Namowitz & N. Spaulding, *Earth Science* (p. 165). Copyright 1985 by D.C. Heath. Reprinted by permission of D.C. Heath and Company.

Figure 4

Young river valleys are V-shaped because while the river is cutting into the valley floor, the valley walls are being worn back by weathering and tributary streams.

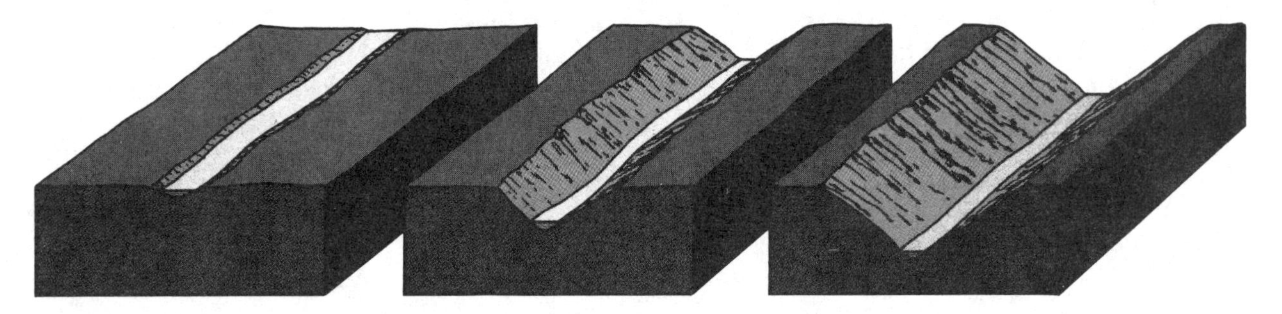

unclear. The publisher included only one photograph which showed a valley after erosion. If a time-lapse series of photographs or drawings had been used, students could see the effects of erosion. For example, the illustrations in Figure 4 help clarify how water erosion cuts river valleys.

Teaching Recommendations

What can teachers do to offset poor text explanations? One approach is to provide students with supplemental materials that present better explanations of important concepts. Schools can purchase several copies of different textbooks, and can subscribe to a variety of popular magazines such as *Current Science, Discover, Outside Magazine, Omni,* and *Naturescope,* all of which contain interesting and well written articles.

A second approach is to encourage more student discussion and writing (see Santa & Havens, this volume). Student-created explanations help students monitor their own understanding and provide the teacher with a better idea of how students view scientific ideas.

In a third approach, teachers can have students create their own visuals. For example, a fifth grade teacher had her students develop their own visuals for a weather unit. The students then used their diagrams to explain concepts to one another in cooperative groups. After the oral presentations, they wrote explanations of their illustrations (Figure 5).

A fourth approach requires teachers to make decisions about reducing coverage of less important concepts. Sometimes teachers have difficulty eliminating content for political or evaluative reasons. One solution is for districts to establish a science committee comprising elementary and secondary teachers. The committee would develop content priorities for each grade level. It is important to limit these priorities so teachers have opportunities to teach for rich conceptual understanding, instead of trying to teach everything.

Finally, the problem of inadequate explanations can be offset by improved practice exercises. Teachers can sort through text exercises and use only those that focus on important concepts. Since most texts contain few thinking tasks, teachers will need to supplement these tasks developing their own questions or by using questions generated by the students.

Eliminating Jargon

Science teachers need to eliminate jargon from their curriculum or at least reduce its importance by focusing student attention on necessary scientific language. The purpose here is not to malign any particular scientific terms but to stimulate debate among science teachers about the appropriateness of the vocabulary in science texts.

Some high school chemistry texts contain an estimated 3,000 words that are unfamiliar to high school students. Compared with the estimated number of words considered adequate

Figure 5
Student Illustration of a Warm Front

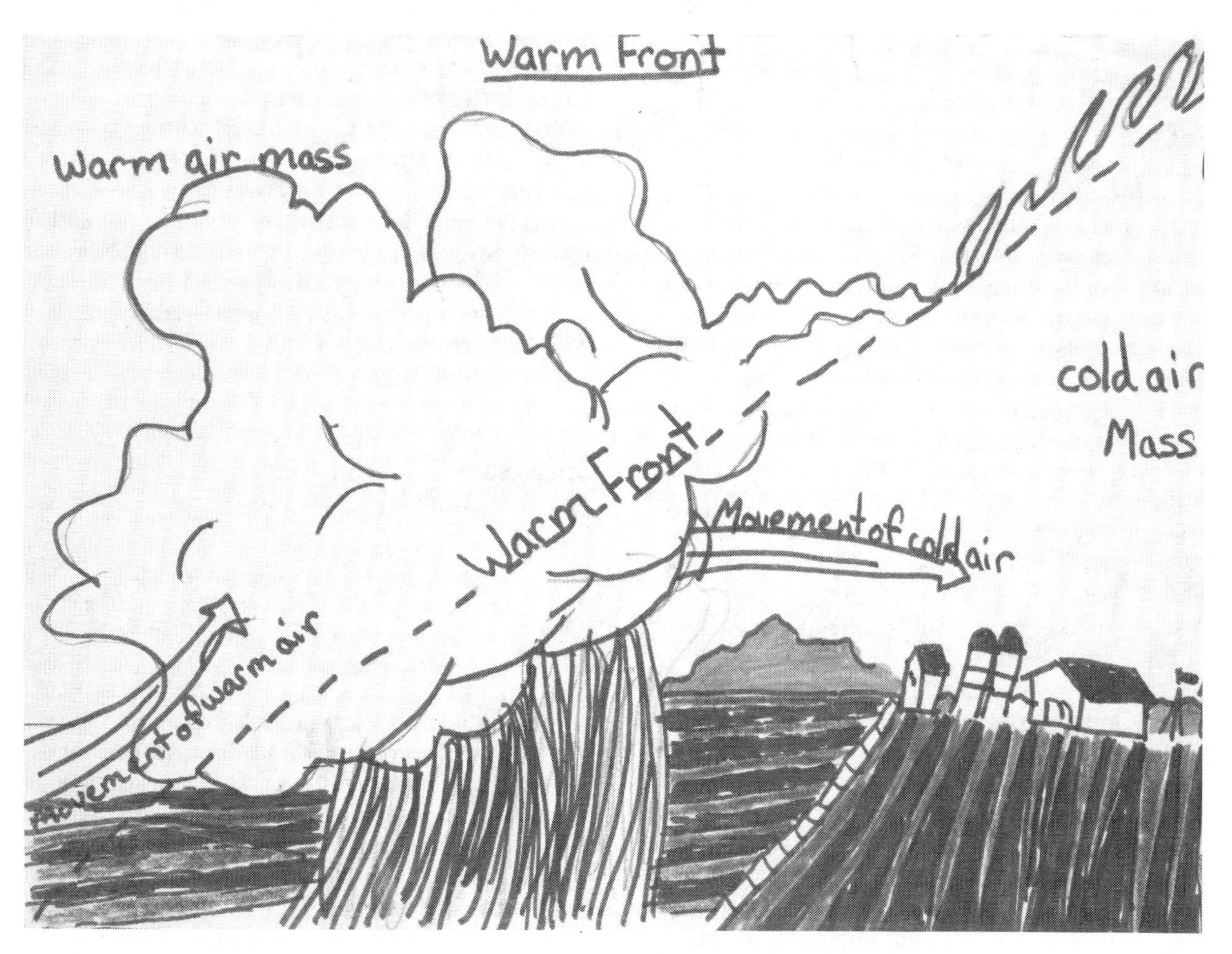

Drawn by Laura Santa

Figure 6
End-of-Chapter Terms from a High School Biology Text

hyphae	rhizoids	conidia	ostiole
mycelium	sporangiophores	asci	apothecia
heterokaryon	sporangia	ascospores	basidiospores
homokaryons	gametangia	perithecia	basidia
stipe	gills	sori	lichen
mycorrhiza	primary mycelium	secondary mycelium	

Modern Biology, Holt, Rinehart & Winston, p. 292, 1985.

for conversational communication (1,500), the number of new words presented in most science textbooks is staggering. It far exceeds the number of words taught in most foreign language classes.

Of course, many of the new words are valuable and useful as cognitive labels. They help students understand scientific concepts, their later coursework, and their interpretation of everyday events. For example, scientifically literate people need to have a good understanding of terms such as chemical equilibrium, force, and plate tectonics. But often we overload students with inappropriate vocabulary (National Research Council, 1990).

Teachers' opinions differ about what to eliminate. Scientific vocabulary can be jargon to students when such words are: (1) used only by experts, (2) difficult to learn (even for high ability students), (3) used only for academic testing purposes, or (4) introduced too early in students' schooling.

Some terms are of questionable relevance to anyone except experts. For example, the list in Figure 6—taken from the end of a chapter on fungi in a widely used high school biology text—contains several highly specialized terms. Students probably should know words like *hyphae* or *heterokaryon,* but why should they be required to learn words like *perithecia* and *ostiole*? These latter terms are unfamiliar to many research biologists, except those dealing directly with botanical topics.

Other words are familiar to scientists but are not necessarily part of the vocabulary of scientifically literate adults. A controversial example is the word *platyhelminthes*, commonly called flatworm. Students need to be exposed to this phylum term, so most biology teachers include it in their curriculum. But why not just mention platyhelminthes in some reasonable fashion and then use the term flatworm? Technical terms often are emphasized and placed on tests without serious consideration or debate by science teachers. Teachers need to decide whether alternative, easier-to-learn terms would be more appropriate for their students.

Another potential problem is that authors often label important concepts and phenomena that probably don't need labels, such as "Le Chatelier's principle" (of dynamic equilibria) or "Bernouli's principle" (fluid pressure). While Le Chatelier appears eight times on a single page in one chemistry text, another popular chemistry text includes an excellent description of the same principle without a single reference to the French discoverer. Similarly, physical science texts emphasize Bernouli's principle as the principle underlying why airplanes stay aloft. Does the label help clarify the concept? The answer is unclear.

Physical scientists do not use these two labels often, but teachers usually do, particularly when testing students' knowledge of the principles. Most scientists describe these phenomena using ordinary words or with vocabulary seldom presented to school students.

Rather than memorizing infrequently used labels, students should learn the characteristics of equilibria and fluid pressures under a wide variety of physical conditions. After all, science is about solving problems under new conditions, not memorizing labels and their definitions in anticipation of exams.

Unfortunately, arguments seldom occur about the appropriateness of traditional terms in texts. Each new text edition contains more pages than the previous edition. New concepts are added without anyone examining whether the old ones should remain.

A final problem is that some terms are presented too early in students' schooling. For instance, a major elementary science series presents the concepts *atrium* and *ventricle,* which are seldom used even by well-educated people. Although these terms need to be learned at some stage of schooling, it should probably be later, in junior high school. Younger students could better spend their time learning functional concepts.

Conclusions

Science texts offer challenges to both teachers and students. Most texts are cluttered with recall questions, inadequate explanations, and scientific jargon. Texts must be improved before they become effective tools for learning science. If we want students to describe and explain natural phenomena, we need materials with questions, explanations, and terminology that facilitate rather than confound this goal.

References

Andre, T. (1979). Does answering high-level questions while reading facilitate productive learning? *Review of Educational Research, 49,* 280-318.

Armbruster, B. (1984). The problem of inconsiderate text. In G. Duffy, L. Roehler, & J. Mason (Eds.), *Comprehension and instruction.* White Plains, NY: Longman.

Heimler, C., & Lockard, J. (1977). *Life science.* Columbus, OH: Merrill.

Herman, P.A. (1984, December). *Incidental learning of word meanings from expository texts that systematically vary text features.* Paper presented at the National Reading Conference, St. Petersburg, FL.

Holliday, W.G. (1990, December). Textbook illustrations. *The Science Teacher, 57,*27-29.

Holliday, W.G., & Benson, G. (1991). Enhancing learning using questions and adjuncts to science charts. *Journal of Research in Science Teaching, 28,* 97-108.

Holliday, W.G., & McGuire, B. (in press). How can comprehension focus students' attention and enhance concept learning of a computer-animated science lesson? *Journal of Research in Science Teaching.*

Holliday, W.G., Wittaker, H.G., & Loose, K.D. (1984). Differential effects of verbal aptitude and study questions on comprehension of science concepts. *Journal of Research on Science Teaching, 21,* 143-150.

National Research Council. (1990). *Fulfilling the promise: Biology education in the nation's schools.* Washington, DC: National Academy Press.

Prawat, R.S. (1989). Promoting access to knowledge, strategy, and deposition in students: A research synthesis. *Review of Educational Research, 59,* 1-41.

Shuell, T. (1990). Phases of meaningful learning. *American Educational Research Journal, 60,* 531-548.

CHAPTER 6 *Reading Science Texts for Conceptual Change*

Kathleen J. Roth

This chapter begins with a description of the importance of teaching students how to achieve conceptual change in science class. Roth challenges us to avoid blaming students for their inability to measure up in class. Students often have rich prior knowledge, but frequently it conflicts with scientific explanations offered in textbooks. Roth takes us into a middle grade classroom for a look at the problems students have reading texts and the reasons their reading strategies work against conceptual change. Then she shows us the process successful readers use to grapple with conflicting concepts. She concludes with ideas about teaching for conceptual change. Elementary teachers will find these ideas adaptable for use with younger children.

Like many middle school science teachers, I was frustrated with science textbooks and students' failures to learn from them. I spent a lot of time and effort working around these texts. Often I abandoned them altogether, providing demonstrations, experiments, discussions, and explanations that I thought made better sense to students. I rationalized that although students were not learning to read science texts, they were developing an understanding of important science concepts and processes. The challenge was to help students develop the same kinds of understanding independently as they read from text.

As a researcher, I wanted to tackle questions about student learning from text. Why are science textbooks so difficult for students to understand? How can teachers help students derive meaning from science texts? But first I had to take a harder look at what students were learning from other modes of instruction. Were the discussions, experiments, and other activities helping students develop the understanding I intended?

A Conceptual Change View of Science Learning

A growing body of research on science teaching and learning provides important insights into why students have learning difficulties in science classrooms. This research clearly demonstrates that despite instruction from enthusiastic, organized teachers, most students are not developing integrated, flexible understanding of science concepts and processes (Anderson & Smith, 1983a, 1983b; Carey, 1986; Champagne, Klopfer, & Anderson, 1980; Johnson & Wellman, 1982; Nussbaum & Novick, 1982b, Roth, 1984). For example, students can memorize definitions and facts about light and how it travels, and they can label the parts of the eye. But they cannot use the memorized definitions and facts to explain other everyday phenomena related to light and seeing (Anderson & Smith, 1983a).

This problem is not simply a matter of student laziness, carelessness, or inattentiveness. Recent research demonstrates that learning new science concepts is a more difficult process than was previously assumed. In the past, students' difficulties with science have been explained by the newness and the abstractness of scientific terms and concepts and by students' lack of prior knowledge about these technical terms (see Finley, this volume). How do you teach about photosynthesis, for example, when students have never heard of the word; have no understanding of chemical reactions, molecules, cells, or chlorophyll; and cannot watch the process happening?

Recent research points to a more powerful explanation for students' learning difficulties than their lack of prior knowledge. Students often have rich prior knowledge about the phenomena they study in science, but frequently that knowledge conflicts with the scientific explanations presented in class and in textbooks. Students may not know the word *photosynthesis*, but they have a lot of ideas related to this concept. They may "know," for example, that plants get their food the way people do—by eating. Plant roots are like mouths and take in food from the soil. This explanation conflicts in critical ways with scientists' understanding about plants' ability to make food internally out of nonfood materials (water, carbon dioxide) taken in from the environment. But the students have constructed these explanations on the basis of their own experiences with plants, and these explanations make sense to them. Personal theories are not easy to give up.

Thus, meaningful learning in science often requires students to go through a difficult process of conceptual change. For this kind of learning to occur, students need to recognize that the scientific concept or explanation conflicts with their own personal theories. They need to be convinced that their own theories are inadequate, incomplete, or inconsistent with experimental evidence, and that the scientific explanations provide a more convincing and powerful alternative to their own notions. Students need repeated opportunities to struggle with the inconsistencies between their own ideas and scientific explanations, to reorganize their ways of thinking, to abandon or modify ideas that have served them well in everyday life, and to make appropriate links between their own ideas and scientific concepts.

Such challenging cognitive and metacognitive work is unlikely to occur in classrooms where students listen passively to a teacher or where students are expected to learn many important concepts and facts quickly. Text materials also can hinder conceptual change. As Holliday notes in the previous chapter, most science texts contain poor questions, inadequate explanations, and too much technical jargon. All of these problems make science learning more difficult. In addition, despite the popular belief that a hands-on approach to teaching is a panacea for students' learning difficulties in science, conceptual change learning does not necessarily occur in activity-focused classrooms. Such learning is best supported by instruction that engages students in actively struggling with the conflicts between their own ideas and scientific explanations in order to construct new understandings.

For the past several years my colleagues and I have been studying the challenges and possibilities of teaching for conceptual change learning in regular science classrooms. One facet of this research has focused on exploring the ways in which students' personal theories (or misconceptions) contribute to their difficulties in learning from science text. The examples presented in the following pages are drawn from this research.

Students' Strategies for Reading Science Text

Thinking about science learning from a conceptual change perspective provides a powerful explanation for why students have trouble reading science text. If students have difficulty learning from classroom instruction, it is not surprising that they have even more trouble learning from independent reading. What happens when students are asked to read texts that conflict with their assumptions about the world? How do they cope with the problems this situation poses?

In a study of middle school students' reading of a textbook chapter on photosynthesis, we identified four reading strategies that failed to promote conceptual change learning. The study provided insights about the reasonableness of these strategies and about how textbooks and teaching practice unintentionally encourage students to develop and use such strategies. The study suggests that teachers need to gain a new appreciation of the challenges students face when told to "Read pages 15-21, define the boldfaced words, and answer the questions on page 22."

With each of the four ineffective strategies, students' incompatible prior knowledge played a critical role; after completing the reading, students continued to use their personal theories to explain everyday phenomena. For example, nearly all the students in the study began and ended the reading believing that plants get their food from the outside environment and that plants have multiple sources of food. But the texts they were reading discussed scientists' understanding that plants' *only* source of food is the internal process of photosynthesis. How did the students miss this message?

Strategy 1: Relying on prior knowledge to complete a school task. Some students relied too heavily on their incorrect experiential knowledge about plants and food in interpreting the text. When asked to recall what the text said, for example, they frequently recounted things that came not from the text but from prior knowledge. Although they reported that the text made sense, these students appeared to avoid thinking about the text as much as possible. If they could decode the words and get enough of the gist of the text to call up appropriate prior knowledge, it made sense. They used this prior knowledge to generate answers to questions at the end of the chapter. These students did not attempt to make sense of text ideas; their goal was to get the reading done and answer the questions. In accomplishing this goal, they tended to ignore text ideas.

Strategy 2: Relying on big words and details to complete a school task. Other students paid much more attention to the text. But the way they paid attention did not help them understand the concepts in the text or relate those concepts to what they already knew. These readers relied too heavily on details in the text, failing to attach any meaning to them. The details—most often specialized vocabulary words—were isolated fragments that students saw as having no relationship to one another or to their real world knowledge. Students felt they understood the text if they were able to pronounce the words and copy appropriate sentences containing the big words to answer end-of-the-chapter questions. (They had learned that using big words can help you get by in school.) They saw this school exercise as having nothing to do with their own ideas about how real plants get food. Their misconceptions remained unchanged, and the big words they had picked up from the text were never used when they talked about real plants. Thus, "school knowledge" in the text was something totally separate from the real world, and it was not expected to make sense or to relate to everyday phenomena.

Strategy 3: Relying on unrelated facts to "learn" science. Another group of students went beyond the focus on big words but still kept their understanding of the ideas presented in the text separate from their knowledge about real plants. These students placed too much emphasis on facts, viewing science learning as a process of developing long lists of facts about natural phenomena. Their prior experiences in school had convinced them that memorizing unrelated facts constitutes satisfactory learning. They accumulated many facts from the text and described these facts accurately. But they treated all the facts as equally important, and they never attempted to relate facts to one another or to their real world knowledge about plants. Reading for conceptual change was therefore impossible. Myra, for example, remembered that the text described photosynthesis as the way plants make their own food. However, when asked to explain how a plant sitting on the windowsill gets its food, she never mentioned photosynthesis or plants' ability to make food.

Strategy 4: Relying on prior knowledge to make sense of text explanations. Another group of students (above-grade-level readers) genuinely tried to make sense of the text and to integrate text ideas with what they already knew about plants. They used a more sophisticated strategy of attempting to link prior knowledge and text knowledge, which is critical for conceptual change learning. However, because their real world knowledge—which was strongly held—often conflicted with the content of the text, these students distorted or ignored some of the text information to make it fit their prior knowledge. They expected the text to confirm what they already knew. They did not read to change what they knew; they read to fill in more details. With prior knowledge in the driver's seat, students' interpretations often were quite different from what was intended by the authors of the text.

Eleven out of 12 students in the study who were reading from commercially available textbooks used one or more of these ineffective strategies during their 3-day reading of the chapter about photosynthesis. These students failed to make sense of the key ideas presented in the chapter and clung to their personal theories about how plants get food.

A critical finding of the study is that for these students, sense making was not a goal in reading science text. They were satisfied if they could successfully complete the school task of generating answers to questions. They reported few occasions when they were actively engaged in puzzling about ideas presented in the text. The students memorized definitions for key words like photosynthesis, and some of them even learned to describe photosynthesis in their own words. But none of the 11 students successfully integrated their entering concepts about how plants get food with the concept of photosynthesis. Thus, they were unable to *use* the concept to explain how real plants get their food, why plants need light, why plants are so important to animals, or whether a plant could live if only some of its leaves received light. Instead, ideas about photosynthesis in the text were treated as isolated, school-bound knowledge.

Strategy 5: Reading for conceptual change. Is it possible for students to use a strategy for reading science text that results in conceptual change? Six out of seven students reading an experimental text about photosynthesis (Roth, 1985) and one of the students reading from a commercial text (Blecha, Gega, & Green, 1979) used such a strategy. These students used the text to change their experiential ideas about how plants get food. They worked hard to integrate concepts presented in the text with their own ideas. They read with the goal of making sense of the text ideas, not just of completing the reading assignment.

In contrast with students who allowed prior knowledge to control their integration of text ideas with their personal theories (Strategy 4), students using the conceptual change strategy allowed text knowledge to take the driver's

TEXT PASSAGE 1

Dependence for food. As you have seen, the roots of flowering plants are adapted to perform certain functions for the plant. They are specially adapted to hold the plant in place and to take in water and certain other materials which the plant needs.

But the materials which the roots take in from the soil are not food. Nor are the roots adapted to make their own food out of these materials. Green plants make food by using the energy in sunlight. But sunlight cannot reach cells that are buried in the ground. For their food, then, the cells of the plant must depend on the other parts of the plant which are above the ground. These other parts, of course, are the stems and leaves.

Stems and leaves. Most of the food produced by flowering plants is produced in their leaves. These leaves are organs that absorb much sunlight. They are made up of specialized cells which enable them to perform their function in the life of the plant.

But how are the leaves connected to the roots? How are they raised and held in the sunlight? This, of course, is a special function of the stems. Within the stems there are cells that have thick, strong walls. Because of these strong cell walls, the stems can hold the leaves up and spread them out in the air and sunlight. In this position, the leaves can manufacture food.

The vascular system. How does food manufactured in the leaves get to the roots? How do the minerals and water absorbed by the roots get up to the leaves? Do these things simply move from cell to cell as water does when it enters the roots?

Within each part of a flowering plant there are certain cells which specialize in the movement of these materials. These cells form bundles of tubelike structures which stretch from the roots, through the stems, and into the leaves. There are two types of these structures. One type, called *xylem* (zī 'lem), carries water and minerals throughout the plant. The other, called *phloem* (flō 'em), carries food.

From *Exploring Science*, Green Book (2nd ed.), ©1979 by Blecha, Gega, Green, & Ide. Reproduced by permission of Macmillan/McGraw-Hill School Publishing Company.

seat in this integration process. They recognized the conflicts between what the text said and their own personal theories and puzzled about these inconsistencies until they could resolve the conflict. Often, this meant changing their personal theories to accommodate scientific explanations.

This kind of processing of text ideas differs greatly from that used in Strategies 1-4. Unlike students using the ineffective reading strategies, the conceptual change readers:

1. Made efforts to link text ideas with their experiential knowledge.
2. Recognized and thought about central text statements that conflicted with personally held ideas.
3. Distinguished between main ideas and supporting details, often minimizing the importance of big words.
4. Experienced and recognized conceptual confusion while reading.
5. Worked to resolve this conceptual confusion.
6. Were aware that their own ideas about real world phenomena were changing.
7. Used concepts presented in the text to explain real world phenomena.

Students who use this sophisticated strategy pay careful attention both to their understanding of the text content (What does that mean? How does it relate to my own ideas?) and to the monitoring of their understanding (Is this confusing to me? How are my ideas changing?). Thus, successful use of the strategy demands challenging cognitive and metacognitive efforts in working toward the goals of making sense of text material.

The Strategies in Action

An analysis of three students' interpretations of sample text passages illustrates why their use of ineffective strategies is reasonable given the way science texts are written and used. It also demonstrates the sophisticated nature of the conceptual change strategy. Text Passages 1 and 2 include scientific explanations that clearly contrast with the students' personal theory that plants take in their food from the soil. Rather lengthy passages are presented so that students' attention to different kinds of text content can be illustrated.

Tracey's use of Strategy 2. A below-grade-level reader (5.6), Tracey was one of the students who focused on the details in the text without making any sense of their meaning. In particular, she focused on italicized vocabulary words, a strategy that enabled her to answer most of the questions at the end of the text chapter. She simply found the big word in the question, looked it up in the text, and copied the sentence. Success with this strategy no doubt reinforced Tracey's understanding that the important parts of the science text were these strange new words and that the purpose of reading was to find these words.

Tracey's summary of Text Passage 1 focused on the section with the most boldfaced vocabulary words—the paragraphs on the vascular system. She remembered the words but nothing about their meaning. When asked if any of the text was confusing for her, Tracey said, "Some words I didn't get." She then opened the book and pointed to the words italicized in the section about the vascular system. She did not report any confusion, however, when she was asked to explain how a real plant gets its food. Despite the text's explicit statement that materials taken from the soil are not food for plants, Tracey described plants' food as water and fertilizers taken in from the soil. Tracey's attention in reading the text was focused almost exclusively on the big words, and since there were no big words in the paragraph on how plants get food, the paragraph did not get Tracey's attention.

In interpreting Text Passage 2, Tracey followed the identical pattern of picking out key vocabulary words without developing any meaning for them. When asked about the main idea of the passage, Tracey replied: "It's about cells—chloro-cells and guard cells." When asked what these cells do, she responded, "Well, they really don't do anything." Tracey again ignored critical paragraphs that conflicted with her own theories. During the interview, for example, Tracey was asked to reread paragraphs 2A-2D and to describe how she would explain that text content to a friend. She summarized the text by looking at it paragraph by paragraph and paraphrasing the first sentence or so of each.

TEXT PASSAGE 2

Production of food. As do all other green plants, flowering plants produce their own food. How do they do this?

Flowering plants, as you know, take in water and certain minerals from the soil. They also take in carbon dioxide from the air. But these materials are not food. No animal, for example, can get energy from them or use them to make the protoplasm of its body. What do flowering plants do to these things to change them into food?

Flowering plants, as do all green plants, use the energy of sunlight to change simple, nonliving materials into food. This food-making process is called *photosynthesis* (fō'te sin'thesis).

Chlorophyll. The chief food-producing organs in the flowering plant are the leaves. Observe the single cell removed from the green leaf in the picture above. Notice that when the cell is greatly magnified it no longer seems to be all green like the leaf. Which part of the cell is actually green?

Notice that the green color is limited to certain small particles in the cell. These particles are called *chloroplasts* (klôr'e plasts). Each chloroplast contains a very complex material called *chlorophyll* (klôr'e fil). It is this chlorophyll that gives green plants their green color. It is this chlorophyll, too, that enables the cells to change the energy of sunlight into the energy stored in food.

Paragraph 2A: Well, plants produce their own food.

Paragraph 2B: Plants take in water and certain minerals from the soil. They have carbon dioxide in the air.

Paragraph 2C: They get energy from the sun.

Paragraph 2D: Chlorophyll is the chief food-producing organ.

This retelling of the text is interesting in several ways. First, Tracey did not mention the key sentences (paragraph 2B) that conflict with her personal theories about plants and food. Second, she continued to focus on big words without attaching any meaning to them. For example, although Tracey had earlier (and mistakenly) described chlorophyll as a cell, here she distorted the text statement and described chlorophyll (again incorrectly) as an organ. Tracey did not see any contradiction because none of these words (chlorophyll, cell, organ) were meaningful to her. Finally, this part of the interview was the only time Tracey made any reference (during three interviews) to the idea that plants make their own food. She never mentioned this concept when asked about real plants, and the only definition she could give for photosynthesis was that it was "some kind of chemical or name or something."

Tracey was not reading to make sense but to complete the school task. Her strategy was to focus on the numerous specialized vocabulary words. Because her attention was focused on finding these words and their definitions (whether they did or didn't have meaning for her), she missed key statements that conflicted with her personal theories about plants and food. Thus, she failed to make connections between the concepts in the text and her own ideas.

Kevin's use of Strategy 4. Kevin was an above-grade-level reader (12.6) who was interested in science and was familiar with at least some of the terminology covered in these text passages. His teacher predicted he would have no trouble with the text. However, Kevin began instruction holding the strong belief that plants have multiple sources of food from the outside environment, and this theory played a major role in his interpretation of the text.

Pretest question: What is food for plants?

Kevin's answer: Food can be sun, rain, light, bugs, oxygen, soil, and even other dead plants. Also warmth or coldness. All plants need at least three or four of these foods.

Kevin attempted to integrate these ideas with the ideas in the text, but he relied so firmly on his prior knowledge that his personal theories changed only slightly and in inappropriate ways to accommodate new text knowledge. He read to fill in the details of what he already knew rather than to make changes in his own theories; this strategy resulted in distortions of text explanations.

Kevin's interpretation of Text Passage 1 illustrates how he ignored or distorted statements in the text to make them fit his own schema of food for plants. Despite the text's statement that water and minerals are not food for plants, Kevin recalled that the passage "told about food from the soil, like minerals and water." He also distorted the section on the vascular system to make it fit with his view that plants get their food from the soil, explaining:

"The picture of xylem and phloem in a plant showed the two layers that make food go through the stem from the soil to the leaves. The xylem takes food from the soil and passes it onto the leaves so it can do photosynthesis. Root hairs...go farther into the soil to get water and minerals and stuff like that, food."

Kevin then described the main idea of the section as being about the "two main parts of food for the plant," saying that the text "would tell what food it would have in the soil and what food it would get from the sunlight."

Thus, he neatly combined his personal notion that food comes from the soil with the text notion of photosynthesis. In his view the top half of the plant near the sun makes its own food and the bottom half of the plant gets food from the soil.

After reading Text Passage 2, Kevin developed a good explanation of photosynthesis, and he recognized that this was an important concept that he needed to fit into his real world schema of food for plants. But again he did not recognize any statements in the text as conflicting with his own view. Even after rereading the paragraph stating that materials taken in from the environment are not food, Kevin did not recognize any conflict. In summarizing how plants get their food, Kevin incorporated photosynthesis into his own theory about multiple sources of food.

Interviewer: So where does this plant get its food?

Kevin: Whew, from lots of places. From the soil for one, for the minerals and the water, and from the air for oxygen. The sunlight for sun, and it would change chemicals to sugar. It sort of makes its own food and gets food from the ground. And from air.

Kevin's strategy represents a fascinating case of how even good readers can severely distort text to fit their prior concept. It is easy to see how a teacher could think that Kevin had made sense of the text view of photosynthesis. If you asked him only, "What is photosynthesis?" you would get an explanation presented in the text. It was only when questions about real plants arose that Kevin's failure to change his prior concept became evident.

Susan's use of Strategy 5. Susan was the only student in the study to use the conceptual change strategy while reading from the commercial text. She used the text to appropriately change her experiential ideas about how plants get their food, successfully integrating information from the text with her real world ideas about plants.

A key to Susan's success was her recognition of statements in the text that conflicted with her own ideas. However, Susan did not pick up this conflict after reading the first text passage, despite the clear statement in paragraph 1B that materials taken in from the soil are not food. Like the rest of the students, she finished the first day of reading continuing to talk about multiple, external sources of food for plants.

Interviewer: What do you think would be food for this plant?

Susan: Well, air, water, sun, and the stuff the cells get...I think it said from the soil...minerals, air, or oxygen or whatever. What also happens at home, when the leaves get all crinkled up and junk, we always give it ...fertilizer.

However, after the second day of reading (Text Passage 2), the first thing Susan described in her recall of the text was an accurate summary of a critical text statement that conflicted with her own ideas (see paragraph 2B): "Well the water isn't food and it gets food out of—I mean it gathers stuff out of the, it gathers water and materials out the soil, and it's *not* food."

Although this statement in the text was not highlighted with bold or italic type, it made a big impression on Susan's thinking. She recognized that the statement was different from her own ideas, and she began to change her ways of thinking about plants' food. Throughout the rest of the interviews she consistently talked about plants getting food only by making it and not by taking it in from the soil. Even when the interview questions were designed to elicit her personal theories, Susan continued to explain phenomena using ideas she had extracted from the text. Thus, she recognized the contradictions between her ideas and those in the text, and she worked through those contradictions until she found the scientific explanation powerful in explaining everyday phenomena.

Interviewer: If I were to cover up all but one leaf of this plant, do you think the way it grows would change?

Susan: Yeah, it would, because there's only one leaf that can change the materials to food and regularly you have much more—and I don't think it could feed the whole plant. I just don't believe it.

Students using the ineffective reading strategies typically answered this question by predicting that the plant would die, explaining their prediction only by referring to personal experiences with plants: the plant would die because it doesn't have enough light. Susan's use of the concept of food-making to explain her prediction reflects a much more sophisticated reading of the text. The text ideas have clearly become meaningful to her, not just memorized facts.

In contrast with Tracey, Susan did not pay much attention to the big words in the text. She ended the unit unsure about the meaning of photosynthesis and its relationship to chlorophyll and chloroplast. Thus, she did not let the technical vocabulary in the text distract her from the meaning and main idea of the text passage.

It is clear from this study that we cannot expect many students to read this kind of text passage and develop a full understanding of the information presented. Even Susan missed the first paragraph that clearly contradicted her personal theories. Given a text that is so full of information that is new to the students, it is not surprising that most of them do not think through the implications of portions that conflict with their own ideas. The language of the text also contributes to this problem, making science seem abstract, foreign, and unrelated to students' own experiences.

Instruction That Supports Reading for Conceptual Change

What can be done to help students like Tracey and Kevin learn how to read such texts for conceptual change? The analysis of students' reading strategies suggests the need to rethink the ways in which science textbooks are typically used in the classroom.

First, it is important to recognize that science textbooks and many common science teaching practices unintentionally encourage students to use the ineffective reading strategies described above. One example, as Holliday noted in the preceding chapter, is asking questions that students can answer by finding the big words in the text and copying the sentences containing those words. Meyer (this volume) refers to this problem in her analysis of the content of elementary science texts. She found that texts consistently emphasized scientific terminology and lacked information linking students' background knowledge to science concepts. As a result, it is reasonable that students would view the text knowledge as separate from their experience. When teachers require students to define all the boldfaced words (some of which may be irrelevant) and to remember lots of details for tests, they force many students to resort to ineffective strategies to cope with the demands of remembering so much new and seemingly meaningless information.

Understanding the limitations of textbooks as aids to conceptual change learning can help teachers better appreciate their students' learning difficulties, but it does not tell them what to do about it. How can teachers encourage students to use a conceptual change strategy for reading science textbooks?

Drawing from research on conceptual change learning in classrooms (Anderson & Smith, 1987; Champagne & Klopfer, 1984; Driver, 1986, 1987; Hewson & Hewson, 1984; Johansson, Marton, & Svensson, 1985; Minstrell, 1984; Nussbaum & Novick, 1982a; Posner, et al., 1982; Roth, Anderson, & Smith,

Table 1
Principles for Promoting Conceptual Change Learning from Text

Common Troublesome Features of Science Textbooks	Resulting Student Difficulties	Ways of Overcoming the Problem and Promoting Conceptual Change Learning
Content coverage is broad and shallow, with emphasis on specialized vocabulary words. Many concepts are covered, but they are addressed superficially, not in ways that promote real understanding.	Encourages students to approach learning science as memorizing lists of unrelated facts and vocabulary words.	Focus on a few critical issues.
Science textbooks are written from scientists' perspectives and do not seriously consider students' ways of thinking.	Students fail to change their ideas, either because they do not see the connections between their own ideas and those in the text or because they distort the text to make it fit their prior knowledge.	Ask questions to elicit and challenge students' thinking and misconceptions.
Teacher's guides for textbooks give correct answers to questions without anticipating or discussing alternative student responses.	Students develop inappropriate strategies for getting right answers and continue to hold critical misunderstandings. They view learning science as a process of getting right answers, even if the answers don't make sense to them.	Probe student responses and give students clear feedback about their ideas.
Explanations of concepts are given in only one way, and explanations of related concepts come in rapid succession.	Students holding different views cannot link the text explanation to their own ideas. They see text explanations as things to be memorized that have nothing to do with their own understandings of the world.	Represent text explanations in different ways that make explicit the contrast and connections between scientific concepts and students' misconceptions.
Activities/experiments are optional supplements that are not closely linked to concepts presented in the text.	Students learn that "doing" science has little to do with reading and thinking about science concepts and that text ideas are separate from the real world. Activities are fun for students but don't help them develop better understandings of concepts.	Select activities to create conceptual conflict and to develop conceptual understanding.
Questions posed to students in textbooks are primarily fact oriented; they rarely require students to construct explanations.	Students think they understand science because they can answer fact questions, but they may continue to hold critical misconceptions that have not been challenged.	Ask questions that give students repeated opportunities to apply text concepts to explain real world phenomena.

1987) and from my study of students' reading strategies (Roth, 1986), I will describe some critical principles of conceptual change instruction. Examples of how each of these principles might be used to help students understand Text Passages 1 and 2 will illustrate how teachers can help students read typical science textbooks using a conceptual change strategy. The principles are summarized in Table 1; several of them are elaborated on below.

Focus on a few critical issues. Science textbooks typically cover many major concepts and focus on specialized vocabulary. As Holliday suggests, teachers should focus on key issues and not try to cover all content.

To use texts to promote conceptual change learning, teachers must use new ways to define the content to be covered. Instead of covering everything in the text, teachers need to identify central concepts that are the most troublesome for students in terms of their conflicting personal theories. The emphasis should be on helping students develop understandings of a few ideas and on enabling them to explore the relationships among these central ideas and their personal theories.

One way of focusing on critical issues is to define a central question or problem that each of the readings and lessons will help to answer. Instead of a broad question (What do you know about plants?) or a question stated in scientific language (What is photosynthesis?), the central question should be a focused one that students can answer at the beginning of the lessons using their personal theories. As the reading and lessons progress, students can use information from the text to change their answer to the question. The central question will keep students' attention focused on the important issues rather than on supporting details that become meaningless in isolation from the overarching concept.

A reasonable central question for Text Passages 1 and 2 is: How do plants get their food? Photosynthesis is the answer, but the question will elicit conflicting student theories. Using this central question and knowledge about students' personal theories, the teacher could select key text passages for emphasis. For example, the paragraphs that directly confront students' misconceptions (1B and 2B) could be highlighted in a reading assignment by asking students to discuss or write about ways in which the information in those paragraphs differs from their own ideas. The section about chlorophyll and chloroplasts might be eliminated from assigned reading or at least postponed until students have developed the main ideas of the passage.

The reading of the section about the vascular system also could be organized around the central question. For example, the teacher could ask students to explain what xylem and phloem have to do with how plants get their food. By continually encouraging students to tie information to this central question, the teacher can help them look for relationships among ideas and develop a meaningful answer to the question. The focus on a central question emphasizes the goal of reading for sense making.

Ask questions to elicit and challenge students' thinking. Textbooks are typically written from scientists' perspectives; in organizing text content, the authors rarely consider students' alternative ways of thinking about topics. Eliciting students' explanations and personal theories throughout instruction can help both the teacher and the student link scientists' explanations with students' ways of thinking. Asking such questions gives the teacher important information about ways in which student thinking differs from concepts in the text. Such questions encourage students to recognize that their ideas are related to (but perhaps different from) what the text is saying. Without such questions, students may fail to consider their own ideas as they read.

In addition to eliciting students' ideas, it is important to listen carefully to their responses, probe their ideas to help them clarify their positions, and respond to their statements in ways that will shape, guide, and support growth in their thinking. By listening for student thinking rather than for right answers, teachers can gather information that will be useful in guiding students to think more carefully about the meaning of the science content.

One way to gather such information is to have students discuss their personal theories and then instruct them to read the text with these theories in mind. Before students read the text passages on how plants get food, for instance, elicit their ideas by having them write answers to the following questions:

> How do plants get their food?
>
> What do the roots of a plant have to do with plants' food?
>
> What do the stems of a plant have to do with plants' food?
>
> What do the leaves of a plant have to do with plants' food?

Collect various students' responses and write a summary of student theories on a classroom chart or overhead transparency. In the discussion that follows, ask questions to get students to clarify their positions. Also ask students to compare and contrast the different theories, challenging them to give evidence to support their theories. Remind students to look for evidence from the text or other sources to support or contradict each of the proposed explanations.

Present text explanations in different ways. Most science textbooks are full of explanations of scientific concepts. However, each concept is usually presented in only one way before the text moves on to explain a related concept. Thus, students confront a rapid-fire sequence of explanations, and do not have time to make meaningful sense of them all. If they did take time, students might find that the way a concept is explained does not make sense given their own naive ways of thinking about it. Unusual students like Kevin, who take time to try to link text explanations to their own understanding of the world, often resolve such confusion by distorting text explanations to make them fit their personal theories. Simply reading textbook explanations once—especially those that conflict with students' everyday ways of thinking about phenomena—is not sufficient to help students like Kevin realize that they need to change their thinking in order to understand the scientific explanation.

To help students understand key explanations in the text, teachers must give students time to grapple with these explanations. In addition, teachers need to help students recognize ways in which the text explanations differ from their own views. Teachers also need to construct different ways of representing the explanation. These representations will be most effective if they make clear the contrasts between students' ideas and scientific explanations.

The key explanations in the photosynthesis text excerpts (Text Passages 1 & 2) focus on the issue of whether minerals and water taken in from the soil constitute food for plants. To allow students to focus on the explanations in these passages, the teacher might first have them look at two critical paragraphs, 1B and 2B. Students could be asked to construct a chart comparing and contrasting the statements in these paragraphs and their own personal theories. This would focus students' attention on the text's explanation that water and minerals are not considered food for plants and on the contrast between this idea and their own views.

The text statements alone, however, may not (and should not) be enough to convince students that their ideas are wrong. The teacher should explore this contrast more fully, getting students to challenge the text explanation by asking, "What further information is needed to make sense of this explanation?" A number of explanations or discussion questions could provide students with alternative representations of this explanation. For example, the teacher could talk about energy-containing food and food that contains no energy. Minerals and fertilizers that plants take in are like vitamin pills people take; they are important, but they are not energy-containing food.

Select activities that create conceptual conflict. The teacher's guides for science textbooks often include suggestions for activities to accompany the reading of the text. It is important to evaluate each activity for its potential contribution to the development of students' conceptual learning. Some activities are fun for students to do, but fail to challenge students' thinking about critical issues. On the other hand, activities or experiments that contradict students' predictions (discrepant events) can provide convincing evidence to students that there may be better explanations for phenomena than their own theories. The discussions surrounding such activities are as important as the activities themselves, for it is through discussion that teachers can help students resolve conceptual conflicts and make sense of the activities.

A common activity teachers use to help students appreciate plants' need for sunlight to make their own food is to grow plants in the light and in the dark. For this activity to really help students understand that water and fertilizers are not food for the plant, it must revolve around the central questions:

> How is the plant in the light getting its food?
>
> How is the plant in the dark getting its food?

The discussion should help students understand the observation that plants in the dark will die even though they are receiving water and fertilizers. If water is food for plants, then why is the plant in the dark dying? The teacher will need to explain that the plants without light are starving to death. Then students should be encouraged to use the concept of photosynthesis to explain why plants in the dark are not getting food but plants in the light are. Does that mean that the sunlight itself is food for the plant? If that is true, then why do plants need water at all? These kinds of student-generated questions could reflect genuine engagement and conceptual conflict.

Give students opportunities to apply text concepts to everyday phenomena. Textbooks pose many questions that require students simply to repeat or recognize facts and definitions. Such questions rarely ask students to construct explanations that connect text concepts with their own experiences. Higher-order textbook questions are likely to be framed as enrichment or optional questions, if they appear at all.

Questions that support conceptual change require students to think about text ideas and to link those ideas to their everyday experiences. Students need repeated opportunities to use their new explanations to explain a variety of real world phenomena. As students see the power of the central concepts to explain many situations, the scientific explanations become more compelling and personally sensible.

To engage students in answering application questions related to the text passages, begin by having them write answers to the following questions:

> When a plant is looking dry and wilted, what do you do to help it?
> Does this mean that water is food for the plant? Explain.
>
> A box is placed over a plant to cover all but one leaf. The plant is watered and has plenty of air, but only that one leaf can receive sunlight. What do you predict will happen? Why?
>
> A drop of rain falls into the soil near the roots of a large bean plant. Describe what will happen to that water if it is taken into the plant.
>
> Look back at what you said about food for plants before reading this chapter. How would you change your statements about how plants get food to make them more accurate and complete?

Then have several students answer the same question aloud, challenging the rest of the class to listen for inaccuracies or incompleteness in the answers. The repetition provides the extra thinking time many students need to come up with their own answers. If only one student answers a given question, the group may move on to the next question before most students have worked through the previous one. Thus, many students are forced to let the faster students do the thinking for them. At the end of the discussion, tell students to revise their written answers.

Difficulties and Dilemmas in Using the Instructional Principles

The principles of conceptual change instruction suggest a great deal of teacher scaffolding of students' experiences with science text. But an important goal is to help students use conceptual change reading strategies independently. How can we help students develop this independence?

It is tempting to think we could just instruct students to pay particular attention to information in the text that conflicts with their own ideas. While this appears to be a reasonable approach, it will not be enough to help most students read for conceptual change. Before students are prepared to do the difficult cognitive and metacognitive work required to take an idea from text and use it to reorganize a long-held personal theory, they must have a notion of what conceptual change entails. They must understand the need to struggle with ideas and to tolerate confusion while working through ideas. In short, they must have meaningful understanding as a goal for their reading of science text. Because many students have as a goal task completion rather than sense making, teachers will need to help them change their goals.

Students will not learn to read for conceptual change overnight. In the early stages, many students need carefully structured experiences that engage them in confronting and resolving the contrasts between their own theories and scientific explanations in their texts. Once students realize the power of this kind of learning and its usefulness in explaining a variety of everyday phenomena, they may find it meaningful to learn explicitly about the strategies needed to achieve such understanding. At this point, teachers may want to discuss the features of the conceptual change reading strategy and help students monitor their use of the strategy while reading text.

To be effective, these principles of instruction require not only careful teaching, planning, and structuring of students' interactions with texts but also knowledge about students' personal theories. How can teachers know which concepts should serve as the focus for a given unit of instruction? How can they learn enough about students' theories to make instructional decisions?

One source of such information is the growing body of research about students' personal theories and misconceptions. This literature includes a number of sources that describe common student misconceptions on a variety of topics in physics, chemistry, biology, and earth science. Finley (this volume) mentions some of his own work along these lines. In addition, many sources describe ways in which students typically misinterpret instruction. Some even describe teaching strategies that help students through the process of conceptual change for a given topic or concept.

Teachers may find it helpful to look at researchers' efforts to incorporate conceptual change teaching strategies into innovative curriculum materials. My colleagues and I have developed materials on numerous topics, including photosynthesis, light and seeing, respiration, and the particulate nature of matter. These materials may be of interest from a reading perspective because they include a student text that is oriented toward conceptual change.

The students themselves are an obvious source of information about students' personal theories. As part of the planning process, teachers can lead a discussion or assign a writing task that elicits and explores students' personal theories. If this discussion is held before the unit is taught, the teacher can use the information elicited to decide on a focus for the unit or reading selection.

Conclusions

If we want to help students learn to read science text for conceptual understanding, we need to consider the difficulties students face in reading science textbooks. The research on conceptual change learning provides new insights into these difficulties. Students' personal theories about natural phenomena complicate the process of reading text, especially when the text provides explanations that conflict with their ideas. To read for conceptual change, students must 1) recognize the similarities and differences between their own ideas and those in the text, 2) struggle with those differences, and 3) reorganize their own conceptual framework to accommodate the scientific explanations. This process requires a sophisticated reading strategy in which students carefully monitor their own thinking and weigh the evidence to decide if a change is warranted. Research on students' reading strategies suggests that most students do not read for conceptual change. Instead, they meet the challenge of reading science text by using a variety of strategies that enable them to complete their school assignments but that do not result in meaningful learning.

Thus, teachers also face a big challenge. They must help students abandon the use of these ineffective strategies and promote the development of a more powerful conceptual change reading strategy. A critical part of this process is helping students develop a new goal for science reading—sense making.

Implementing these conceptual change instructional principles will be a time-consuming and difficult task. Meaningful use of the principles depends on teachers having a deep understanding of both science and their students' thinking, as well as genuine respect for the challenges students face in undergoing conceptual change. Is it worth the effort? Research evidence supports the effectiveness of these principles in helping students understand text content. It remains an open, empirical question, however, whether such teaching over time can help students develop a conceptual change strategy they can use in reading text independently. Meeting this challenge will require patient and consistent work with students over time to help them develop sense making as the major goal of science learning and reading.

References

Anderson, C.W., & Smith, E.L. (1983a). *Children's conceptions of light and color: Developing the concept of unseen rays.* Paper presented at the annual meeting of the American Educational Research Association, Montreal, Canada.

Anderson, C.W., & Smith, E.L. (1983b). *Teacher behavior associated with conceptual learning in science.* Paper presented at the annual meeting of the American Educational Research Association, Montreal, Canada.

Anderson, C.W., & Smith, E.L. (1987). Teaching science. In V. Richardson-Koehler (Ed.), *Educators' handbook: A research perspective.* White Plains, NY: Longman.

Blecha, M.K., Gega, P.C., & Green, M. (1979). *Exploring science, green book* (2nd ed.). River Forest, IL: Laidlaw.

Carey, S. (1986). Cognitive science and science education. *American Psychologist, 41,* 1123-1130.

Champagne, A.B., & Klopfer, L.E. (1984). Research in science education: The cognitive psychology perspective. Reprinted from D. Holdzkom & P.B. Lutz (Eds.), *Research within reach: Science education* (pp. 171-189). Charleston, WV: Appalachia Educational Laboratory, Research and Development Interpretation Service.

Champagne, A.B., Klopfer, L.E., & Anderson, J.H. (1980). Factors influencing the learning of classical mechanics. *American Journal of Physics, 48,* 1074-1079.

Driver, R. (1986). *Restructuring the physics curriculum: Some implications of studies on learning for curriculum development.* Paper presented at the International Conference on Trends in Physics Education, Tokyo, Japan.

Driver, R. (1987). Promoting conceptual change in classroom settings: The experience of the Children's Learning in Science Project. In. J.D. Novak (Ed.), *Proceedings of the second international seminar on misconceptions and educational strategies in science and mathematics.* Ithaca, NY: Cornell University.

Hewson, P.W., & Hewson, M.G. (1984). The role of conceptual conflict in conceptual change and the design of science instruction. *Instructional Science, 13,* 1-13.

Johansson, B., Marton, F., & Svensson, L. (1985). An approach to describing learning as change between qualitatively different conceptions. In L.H.T. West & A.L. Pines (Eds.), *Cognitive structure and conceptual change.* New York: Academic.

Johnson, C.N., & Wellman, H.M. (1982). Children's developing conceptions of the mind and the brain. *Child Development, 53,* 222-234.

Minstrell, J. (1984). Teaching for the understanding of ideas: Forces on moving objects. In C.W. Anderson (Ed.), *Observing science classrooms: Perspectives from research and practice.* Columbus, OH: ERIC Center for Science, Mathematics, and Environmental Education.

Nussbaum, J., & Novick, S. (1982a). Alternative frameworks, conceptual conflict, and accommodation: Toward a principled teaching strategy. *Instructional Science, 11*(3), 183-200.

Nussbaum, J., & Novick, S. (1982b). *A study of conceptual change in the classroom.* Paper presented at the annual meeting of the National Association for Research in Science Teaching, Chicago, IL.

Posner, G.J., Strike, K.A., Hewson, P.W. & Gertzog, W.A. (1982). Accommodation of a scientific conception: Toward a theory of conceptual change. *Science Education, 66*(2), 211-227.

Roth, K.J. (1984). Using classroom observations to improve science teaching and curriculum materials. In C.W. Anderson (Ed.), *Observing science classrooms: Perspectives from research and practice.* Columbus, OH: ERIC Center for Science, Mathematics, and Environmental Education.

Roth, K.J. (1985). *Food for plants: Teacher's guide* (Research Series No. 153). East Lansing, MI: Michigan State University, Institute for Research on Teaching.

Roth, K.J. (1986). *Conceptual-change learning and student processing of science texts* (Research Series No. 167). East Lansing, MI: Michigan State University, Institute for Research on Teaching.

Roth, K.J., Anderson, C.W., & Smith, E.L. (1987). Curriculum materials, teacher talk, and student learning: Case studies in fifth-grade science teaching. *Journal of Curriculum Studies, 19*(6), 527-548.

CHAPTER 7

Understanding Science Text and the Physical World

Audrey B. Champagne
Leopold E. Klopfer

This chapter demonstrates how misconceptions can interfere with students' understanding of force and motion. Because misconceptions are difficult to change with traditional science learning activities, Champagne and Klopfer recommend alternative approaches that engage students in social interactions with their peers and their teachers. Students need to interact to clarify and elaborate on their views before, during, and after reading. The authors lead us through an interactive mapping strategy called ConSAT *and show us how the strategy works to induce conceptual change. Although the authors use examples from secondary school science texts, the strategy can be easily adapted for upper elementary use.*

The knowledge students bring to an academic task such as comprehending text or observing and explaining physical phenomena largely determines what they remember and observe. For instance, research has demonstrated that students' belief that the rate at which an object falls is proportional to the object's weight influences both what they remember from reading and their observations of falling objects. If you try the experiment outlined in Figure 1, you will see how this influence operates.

After the experiment, if you ask students about the information or experiences they applied in making their predictions, you may discover that many of them referred to a thought experiment proposed by Galileo to support his contention that vacuums are not a natural condition. This experiment, recounted in many science texts, involves dropping a feather and a gold coin in air and in a vacuum. Galileo observed that in air the coin fell faster than the feather, but asserted that in a vacuum both would fall at the same rate. (This mode of argumentation is called a thought experiment because one premise of the argument cannot be verified empirically.)

Students' use of Galileo's experiment to support their predictions demonstrates how their preconceptions influence what they un-

Figure 1
Classroom Experiment on Falling Objects

1. Locate two cubes of the same size, each approximately 5cm on a side. One cube should be made of aluminum and the other of plastic.
2. Hold the cubes one meter above the floor and tell students that you will release them simultaneously.
3. Ask students to predict how the times for the two blocks to reach the floor will compare. Have them give the reasons for their predictions.
4. Drop the cubes several times so students can observe them fall.

If your students are like the ones in our experiment, most of them (over 80 percent) will predict that the aluminum cube will fall faster than the plastic cube. Students who make this prediction usually either report observing the aluminum cube fall faster or criticize the design of the demonstration.

Explanation. Under the conditions of the demonstration (objects of the same size and shape falling a small distance), no difference can be observed in the time it takes the two cubes to fall. Thus, students' observation that the aluminum cube falls faster is neither objectively correct nor consistent with Newtonian theory. This observation is consistent, however, with students' previous observation of objects in motion.

The results of this and similar experiments show that the knowledge a person brings to a science demonstration influences what is observed. This effect is particularly evident when this knowledge conflicts with the objective observation. Under these conditions, seeing is not believing—believing is seeing.

derstand and recall from text. The students say their prediction that the metal cube will land first is based on Galileo's *proof* that heavier objects fall faster. They correctly recall that Galileo's experiment involved dropping a feather and a coin under normal conditions (in air) and that the coin fell faster. They do not recall the thought experiment part in which the objects are dropped in a vacuum; nor do they remember Galileo's prediction about the results of such an experiment. On the basis of what they recall, the students conclude that Galileo proved that heavy objects fall faster—a conclusion consistent with their basic belief that speed is proportional to weight.

These examples demonstrate the influence of students' prior knowledge on what is recalled from text or observed about a physical event. In general, studies of text comprehension indicate the helpful effect of world knowledge. As our research on science learning indicates, however, when students' world knowledge conflicts with the science concept under study, learning is hindered (see Finley and Roth, this volume). As a consequence, traditional science learning activities often fail to change nonscientific world knowledge.

There is ample empirical evidence that traditional methods of teaching science are not effective in this regard. Sociopsychological theory, however, has implications for alternative instructional methods. These newer techniques may be more effective than traditional methods in overcoming the tendency of existing knowledge to inhibit conceptual change. These methods engage students in social interactions with their peers and teachers in a way that encourages development of the cognitive skills necessary for analyzing the natural world and comprehending science text.

Social Interaction and Cognitive Change

Largely as a result of Vygotsky's (1962) influence, the idea that social interaction is a useful instructional tool has grown in popularity and has been applied to many subjects across all grades. The results reported are generally positive. They include improved reading achievement, fuller understanding of concepts, increased awareness of one's own learning processes, raised cognitive level, and improved problem-solving capabilities. When students with different ideas are asked to reach a consensus through cooperative social interaction, teachers have reported increased achievement, retention, and critical reasoning skills.

Social interaction produces two types of conceptual change: the development of concept understanding and the development of cognitive skills such as elaboration. Elaboration involves adding information to existing knowledge. The information may be either declarative (knowing *what*) or procedural (knowing *how*). Elaboration of an academic task involves searching memory for relevant declarative knowledge and procedures and then generating a plan to complete the task. Elaboration results in a well-structured statement based on valid information. For elaboration to occur, the learner must be actively engaged in the academic task.

Active engagement in a science task can be achieved by describing a demonstration (e.g., dropping an aluminum and a plastic cube simultaneously) and then asking students to predict the outcome and to provide a rationale for their prediction. To complete this activity, students reflect on their understanding of the task, generate a plan for accomplishing it, recall pertinent information, make a prediction, and develop explanations for the concept behind the proposed demonstration. Students thus bring personal information to the task.

Prediction and explanation are important to students' active participation in group discussions. Presumably, students have developed some commitment to the product of their thinking and will be motivated to share these ideas with other students and to defend the wisdom of their thinking. These peer interactions, moderated by the teacher, help students augment the quantity and quality of the information they apply to the task, as well as to strengthen the structure of the explanations they have developed. The group's task is to develop a consensus; consequently, each student must listen carefully to the presentations of others in the group. This exposes students to new information and alternative ways of thinking about the task, and enables them to modify their ideas accordingly.

Inevitably, controversies arise over the validity of information and the logic of explanations. The teacher (as coach and referee) is responsible for helping the group recognize the need for standards, as well as for helping define procedures to assess the validity of information and the logic of explanations. The teacher also must see that the procedures are applied, while at the same time helping the group take increased responsibility for the monitoring function. Procedures that are applied overtly in the discussions among students become internalized as cognitive skills.

The skills learned through elaboration are analogous to the skills competent readers apply to comprehending text. Having ascertained that the purpose of a science text is to explain a physical process, competent readers determine what they already know about the process and their explanation for it. As they read, they compare their knowledge with that presented in the text. They also assess the text's evidence and the logical quality of the explanation. Competent readers do not passively absorb the information presented. Rather, they interact with the text, checking their information about the topic against that in the text and applying standards for evidence and logical reasoning. We know of no empirical evidence demonstrating the validity of our conjecture about a parallel between elaboration skills learned in social interactions and the skills applied in comprehending science text; however, this conjecture is consistent in many respects with the empirically tested ideas of Palincsar and Brown (1984).

Integration Using ConSAT

Another useful instructional method is to focus on integration, or the process by which the structural organization of information is discovered. The Concept Structure Analysis Technique (ConSAT) is a concept mapping strategy that can help students develop proficiency in this cognitive process. One application of ConSAT is to guide students in comprehending difficult science text—especially text containing ideas that conflict with students' world knowledge.

Understanding Force and Motion

ConSAT can be used to develop students' skills in representing text structure. It also illustrates how elaboration, stimulated by students' social interactions in the classroom, contributes to comprehension of text. To gain an understanding of how the strategy works, read the Text Passage on the next page, which is typical of material physics students might be assigned to read.

To summarize, the passage is about two kinds of reasoning—intuitive and scientific—and how these lead to two different views of the relationship between force and motion. Understanding the difference between these types of reasoning requires understanding how Aristotle's and Galileo's views of motion differ. Physics teachers know that students will have a difficult time distinguishing between these views. Aristotle's intuitive view matches the relationship between force and motion that most students believe is true. Consequently, students understand and remember the Aristotelian view, while they tend to forget the view proposed by Galileo.

On the basis of research, we can predict that students' interpretation of the text will be influenced by their belief in the principle that motion is proportional to force. Several variations on this theme exist: forces cause motion, no motion means no force, rapid motion means a large force, slow motion means a small force. The idea that motion is proportional to force will be evident in students' interpretations of the relationship between force and motion in the three situations described in the passage—a body at rest, a carriage drawn by horses, and a cart moving along a level road. Where the body is at rest, there is no motion, so students will conclude there is no force. Where the force on the carriage is doubled from two to four horses, students will conclude that the motion or speed also doubles.

The third case leads to the undoing of the intuitive principle. In the real world situation, there is friction in the wheels, friction between the wheels and the road, and friction produced by the air pushing against the cart. Under these conditions, a constant forward push keeps the cart moving at a constant speed. When the forward push stops, the frictional forces slow the cart until it stops. Were it possible to make the wheels frictionless and to move the cart in a vacuum, the cart would accelerate as long as the force persisted and would continue to move at a constant speed when the force stopped.

The idealized case of the cart moving in a vacuum on a level, frictionless road contradicts the principle that motion implies force. Aristotle and most students share the belief that force causes or is proportional to motion. Galileo substituted the principle that forces change motion. Aristotle's interpretation of

TEXT PASSAGE

A most fundamental problem, for thousands of years wholly obscured by its complications, is that of motion. All those motions we observe in nature, that of a stone thrown into the air, a ship sailing the sea, a cart pushed along the street, are in reality very intricate. To understand these phenomena it is wise to begin with the simplest possible cases, and proceed gradually to the more complicated ones. Consider a body at rest, where there is no motion at all. To change the position of such a body it is necessary to exert some influence upon it, to push it or lift it, or let other bodies, such as horses or steam engines, act upon it. Our intuitive idea is that motion is connected with the acts of pushing, lifting or pulling. Repeated experience would make us risk the further statement that we must push harder if we wish to move the body faster. It seems natural to conclude that the stronger the action exerted on a body, the greater will be its speed. A four-horse carriage goes faster than a carriage drawn by only two horses. Intuition thus tells us that speed is essentially connected with action.

It is a familiar fact to readers of detective fiction that a false clew muddles the story and postpones the solution. The method of reasoning dictated by intuition was wrong and led to false ideas of motion which were held for centuries. Aristotle's great authority throughout Europe was perhaps the chief reason for the long belief in this intuitive idea. We read in the *Mechanics,* for two thousand years attributed to him:

> The moving body comes to a standstill when the force which pushes it along can no longer so act as to push it.

The discovery and use of scientific reasoning by Galileo was one of the most important achievements in the history of human thought, and marks the real beginning of physics. This discovery taught us that intuitive conclusions based on immediate observation are not always to be trusted, for they sometimes lead to the wrong clews.

But where does intuition go wrong? Can it possibly be wrong to say that a carriage drawn by four horses must travel faster than one drawn by only two?

Let us examine the fundamental facts of motion more closely, starting with simple everyday experiences familiar to mankind since the beginning of civilization and gained in the hard struggle for existence.

Suppose that someone going along a level road with a pushcart suddenly stops pushing. The cart will go on moving for a short distance before coming to rest. We ask: how is it possible to increase this distance? There are various ways, such as oiling the wheels, and making the road very smooth. The more easily the wheels turn, and the smoother the road, the longer the cart will go on moving. And just what has been done by the oiling and smoothing? Only this: the external influences have been made smaller. The effect of what is called friction has been diminished, both in the wheels and between the wheels and the road. This is already a theoretical interpretation of the observable evidence, an interpretation which is, in fact, arbitrary. One significant step further and we shall have the right clew. Imagine a road perfectly smooth, and wheels with no friction at all. Then there would be nothing to stop the cart, so that it would run forever. This conclusion is reached only by thinking of an idealized experiment, which can never be actually performed, since it is impossible to eliminate all external influences. The idealized experiment shows the clew which really formed the foundation of the mechanics of motion.

Comparing the two methods of approaching the problem we can say: the intuitive idea is—the greater the action the greater the velocity. Thus the velocity shows whether or not external forces are acting on a body. The new clew found by Galileo is: if a body is neither pushed, pulled, nor acted on in any other way, or, more briefly, if no external forces act on a body, it moves uniformly, that is, always with the same velocity along a straight line. Thus, the velocity does not show whether or not external forces are acting on a body. Galileo's conclusion, the correct one, was formulated a generation later by Newton as the *law of inertia.* It is usually the first thing about physics which we learn by heart in school, and some of us may remember it:

Every body perseveres in its state of rest, or of uniform motion in a right line, unless it is compelled to change that state by forces impressed thereon.

A. Einstein and L. Infeld, *The Evolution of Physics*, Simon and Schuster, 1938. Permission to reprint granted by the Albert Einstein Archives, Hebrew University of Jerusalem.

force and motion answers the question, "Why do objects move?" Galileo's analysis answers the question, "Why do objects change motion?" Aristotle's view is based on his observations of real carts; Galileo's is based on an idealized cart on an idealized road. Aristotle's approach is intuitive, Galileo's idealized.

A more basic point can be understood by looking at the nature of Galileo's contribution to the discussion, which caused scientists to change their way of thinking about motion. Galileo did not trust Aristotle's intuitive notion that because force and motion are associated in our experience, we should conclude that there can be no motion without force. He asserted that it is possible for a body to move even when no force acts on it. This important change in the thinking process, basically a change from intuitive reasoning to what today is called scientific reasoning, contributed directly to Isaac Newton's development of classical mechanics. Einstein and Infeld's (1938) main message in this passage is that intuitive conclusions based on immediate observation are not always to be trusted.

Using ConSAT in the Classroom

Teachers assigning this text would like their physics students to understand the main message and to be able to characterize the Aristotelian and Newtonian views of motion. After their study of the passage, students should understand the distinction between intuitive and scientific reasoning and the difference between Aristotle's and Galileo's views of the relationship between force and motion. To help students attain this understanding, a teacher might begin by showing them how to use ConSAT to generate individual concept maps. The teacher could then use social interaction techniques to help students elaborate and achieve a consensus on the concept structures that best represent the information in the passage.

Before introducing ConSAT to your students, you will want to choose a science passage that is appropriate for their reading ability and grade level. The procedure described below can be adapted for use with whatever passage you choose.

First, tell the students what the passage is about (the two kinds of reasoning and how they lead to different views of the relationship between force and motion). Before asking students to read the passage, help them come to a consensus about their understanding of the key terms (force and motion) and the relationship between them. To engage students in the consensus-building activity, follow the ConSAT procedure outlined here:

1. Before reading the passage, students are given large sheets of paper and asked to write the concepts and phrases they think of in connection with motion, and to arrange these concepts in a diagram showing how they are linked. On the same sheet of paper, students draw a second diagram showing the link between the concepts and phrases they associate with force. Finally, students draw a diagram to show the concepts and phrases they use to describe the relationship between motion and force. When all students have made their three representations, the teacher uses the social interaction skills previously described to help the class reach a consensus on the single structure representing the group's current understanding of motion and force and the relationship between them.

2. Next, the students read the text passage to find out what it says about these key concepts and their relationship. The teacher asks

Figure 2
Class Representation of Text: Description of Motion and Force

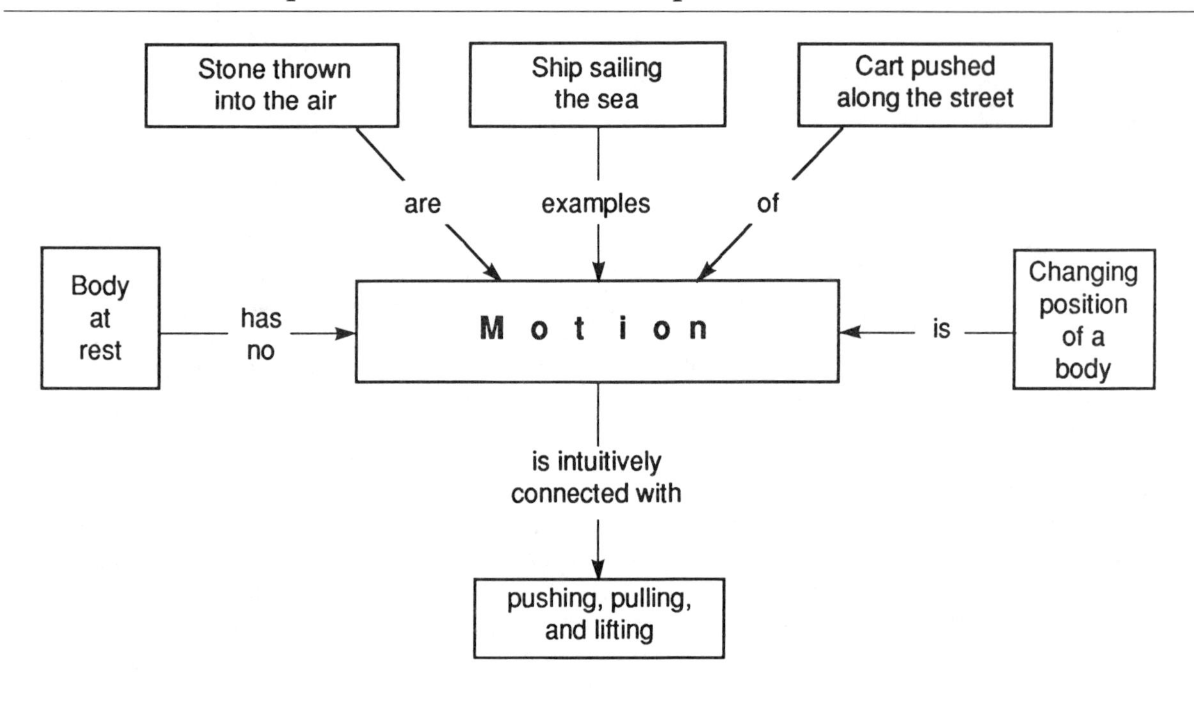

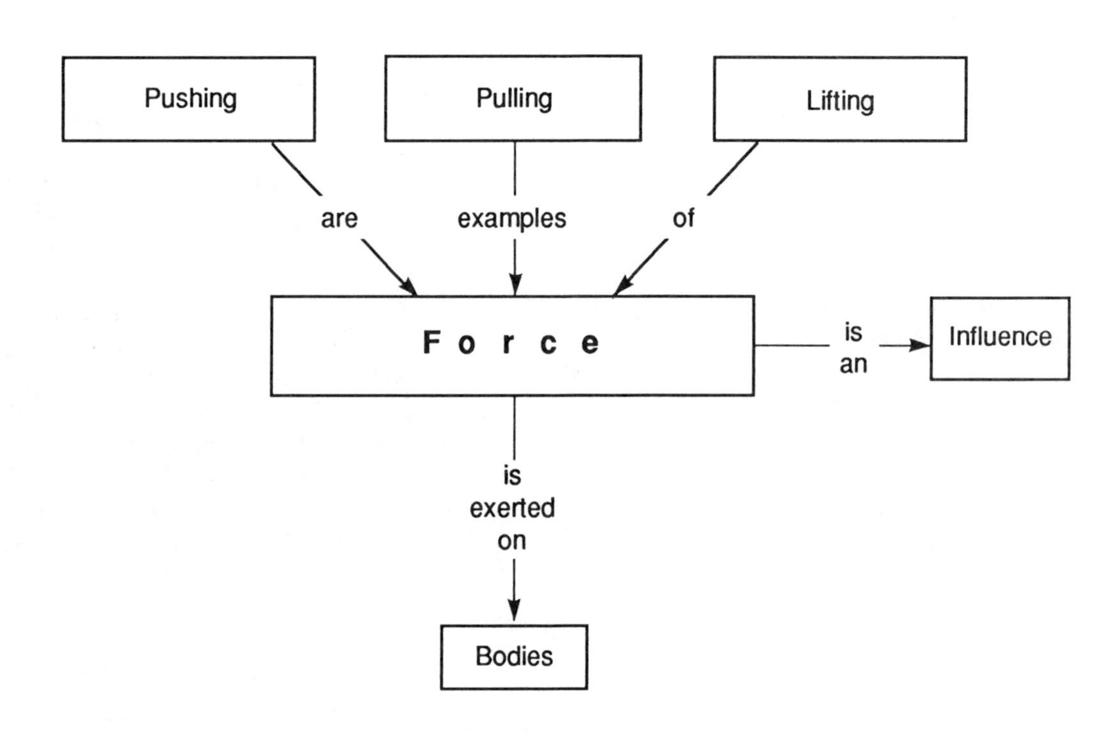

Figure 3
Class Representations of Aristotelian and Galilean-Newtonian Views of Motion and Force

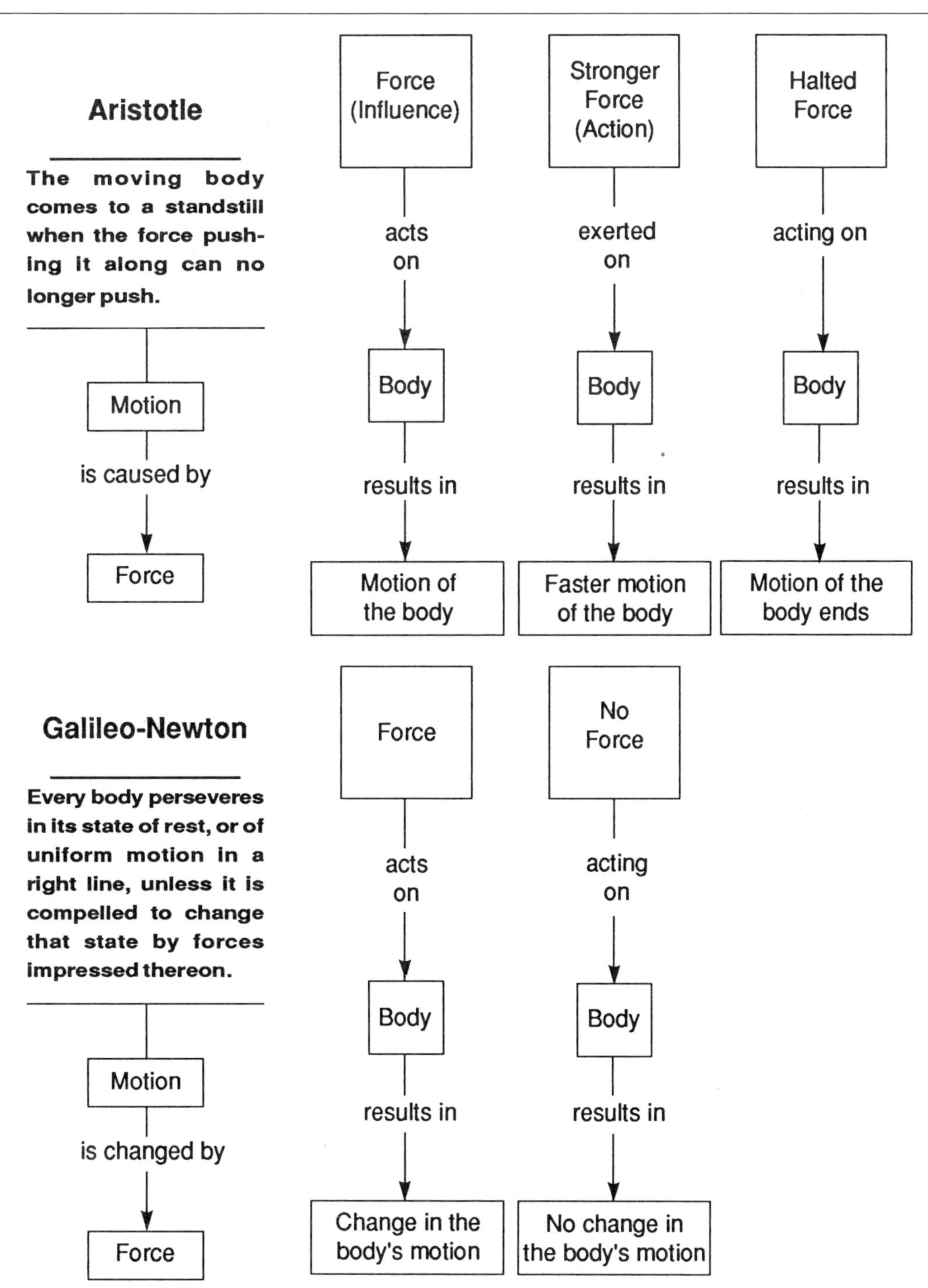

Figure 4
Class Representations of Intuitive and Scientific Reasoning

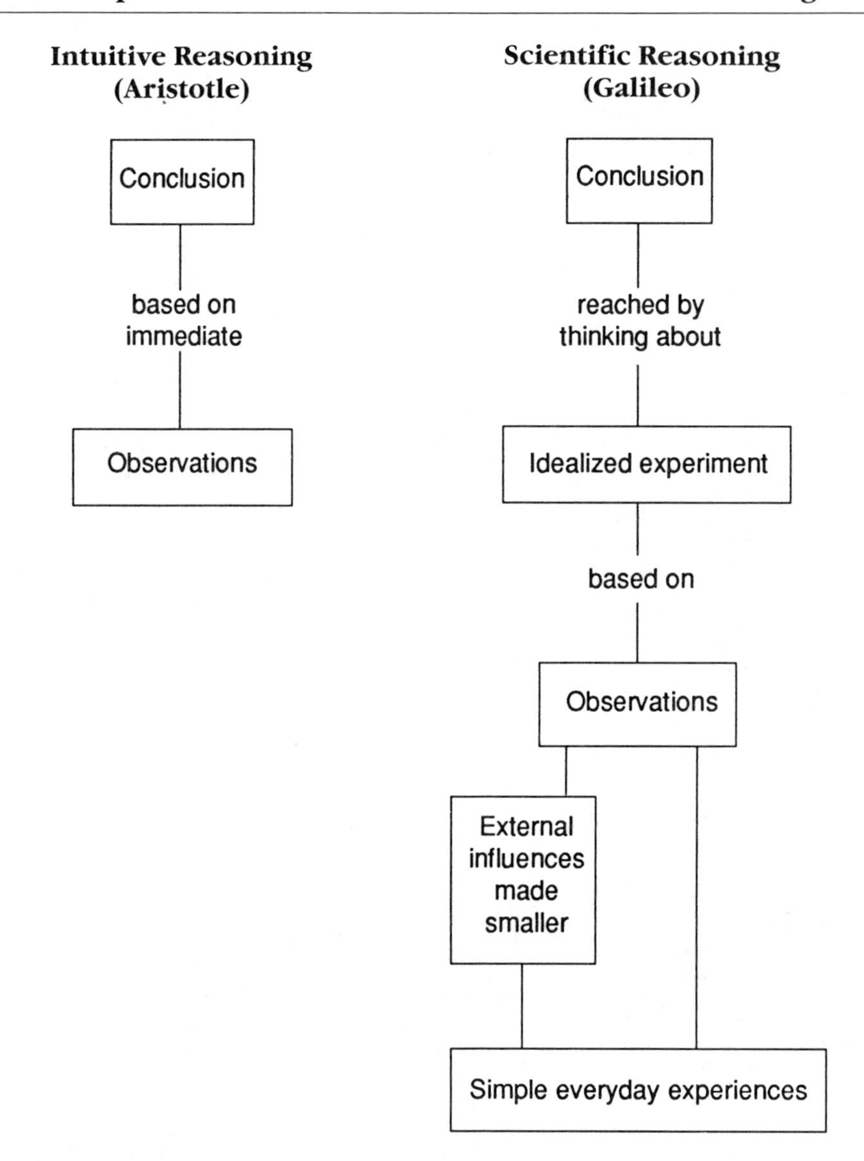

students to draw individual diagrams to show their interpretation of the words, phrases, and links the authors use to describe motion. Pointing out that the authors sometimes use the word influence as a synonym for force, the teacher asks students to draw a second diagram to show their interpretation of the words, phrases, and links the authors use to describe force. When the students finish, the teacher brings them together and helps the class work toward developing a representation everyone can agree on. The resulting representations of the authors' descriptions of motion and force might look like those in Figure 2.

3. The students now go back to the passage and individually prepare two more structural representations. One is a diagram of the words and phrases Aristotle used to describe the relationship between force and motion, and the second is a diagram of that relationship in the words and phrases used by Galileo and Newton. The teacher then assembles the group and guides students' efforts in agreeing on a single representation of the two contrasting relationships. The group representation could be something like the one in Figure 3.

Students' efforts to reach a consensus on the representations of their existing knowledge and of the information presented in the passage force comparisons between the representations and make the differences explicit. This procedure reduces the chance that existing knowledge will dominate students' interpretation of the relationships in the passage.

4. Having guided the students toward recognizing the differences between the Aristotelian and Newtonian views, the teacher will want the students to be aware of the text passage's main message (the distinction between intuitive and scientific reasoning). The same procedures are used as before. Students are asked to reread the text for information about intuitive and scientific reasoning. They individually draw two structural diagrams, one showing the words and phrases used to describe intuitive reasoning and the other showing the words and phrases that describe scientific reasoning. Then the group works together toward an agreed on representation, such as the one illustrated in Figure 4.

When the class has reached a consensus on a representation, individual students will have developed an understanding of the passage. In addition, they will have used external dialogue to question text and to match their understanding of the topic against the author's. Internalizing these dialogues is at the heart of understanding both science text and the physical world.

References

Einstein, A., & Infeld, L. (1938). *The evolution of physics.* New York: Simon & Schuster.

Palincsar, A.S., & Brown, A.L. (1984). Reciprocal teaching of comprehension-fostering and comprehension-monitoring activities. *Cognition and Instruction, 1,* 117-175.

Vygotsky, L. (1962). *Thought and language.* Cambridge, MA: MIT Press. (Original publication, *Myshlenie i rech,* Moscow, 1934.)

PART 3

Classroom Approaches

The two chapters in this section focus on classroom approaches to teaching and learning science. When reading these chapters, imagine a continuum with text-bound experience at one end and hands-on experience at the other. Located between these endpoints are a number of approaches that use print and nonprint materials in various ways.

Although both chapters focus on secondary school teachers and students, several of the strategies involving predicting and critical thinking are equally appropriate for use in elementary classrooms. In Padak and Davidson's chapter, elementary school teachers will find the sample problem solving lesson an excellent way to develop younger students' ability to make and verify predictions. In Alvermann and Hinchman's chapter, elementary teachers will identify with some of the same problems secondary school teachers experience in helping students think critically about difficult science concepts.

This section on different classroom approaches to teaching and learning science promises to stimulate thinking about ways to improve science instruction in classrooms at all grade levels.

CHAPTER 8

Instructional Activities for Comprehending Science Texts

Nancy D. Padak
Jane L. Davidson

Padak and Davidson describe the differences in student learning observed in their research on teacher/manual-generated lessons and teacher/student-generated lessons. They demonstrate through actual classroom dialogue how different approaches to teaching and learning from science text affect students' level of understanding.

Over the past several years, we have examined lessons in which teachers and students explore science concepts through reading and discussion. This chapter provides excerpts from two types of discussions about science text: a "business as usual" approach—involving the teacher's interpretation of suggestions for discussion as found in a teacher's manual—and a problem solving approach. Differences in approaches influence students' understanding of text concepts. Students who participate in problem solving discussions are more likely to understand major text concepts than are students who participate in business as usual discussions (Wilkerson, 1984, 1986).

Our examples are taken from a unit about light and color in an eighth-grade general science text. Some key concepts included in this unit are spectrum, electromagnetic spectrum, diffraction, reflection, refraction, prisms, and concave and convex lenses. Because these ideas are complex, discussion is essential.

Beginning the Reading

The lesson begins with eighth grade students exploring how to hold prisms to the sunlight so the spectrum appears on a piece of white paper. Students perform this experiment, observe the results, and share their experiences with the group prior to reading. The text assigned begins with an introductory paragraph outlining the spectrum experiment, and then goes on to discuss light and color. For comparison purposes, a business as usual lesson and a problem solving lesson involving the assigned text follow.

Business As Usual Lesson

Teacher: I'd like to have you read the first section. Start on page 396, continue on the top of page 397, and stop at the end of that paragraph. Everybody got it? (Students read the introductory paragraph and the first paragraph of the section on light and color.)

Problem Solving Lesson

Teacher: I'd like you to begin by reading the first paragraph. Then tell me what you expect to find in this passage. (Students read the introductory paragraph.) You know it's about light and color and spectrums. What else do you expect to find?

Robin: Heat.

Teacher: Why do you say *heat*?

Robin: Well, some of the colors are cooler.

Teacher: Do you know which ones are cooler?

Robin: I think the darker ones.

Teacher: Why?

Robin: They look cooler.

Teacher: They look cooler. OK. What else? Do you agree or disagree?

Don: Well, I agree with her on infrared and ultraviolet. They are probably the hottest colors you can get in the spectrum.

Teacher: All right. Anybody else?

Sam: Because whenever you melt steel, it always turns red before it turns white and melts completely.

Teacher: OK.

John: You can't see infrared.

Teacher: How would that make a difference in what Sam just said?

John: Well, he just said that it turned red before it turned white. And really, you can't see white. It's just a shade. Infrared you can't see—which would be just like sunlight. So...I think it would be hotter.

Teacher: You think it would be hotter. OK. Anybody else? (pause) I want you to read the paragraph beginning on the bottom of the page and ending on

Business As Usual Lesson	Problem Solving Lesson
	the top of the next page. Cover up what's below it with your paper. Read that far and then stop. (Students read the first paragraph of the section on light and color.)

In both lessons, the teacher began by providing the students with a concrete demonstration of the concept to be studied. By manipulating the prism and observing its effect on light, the students saw how light can be broken into colors that correspond to different wavelengths. In addition, students shared with one another their perceptions of the demonstration.

Beyond this, however, there were dramatic differences in the way each lesson began. In the business as usual lesson, the teacher simply assigned a portion of the text and asked students to read it with no reason provided. Presumably, students read because the teacher told them to. During interviews we conducted the day after this lesson, several students mentioned that during reading they thought about questions the teacher might ask, and they read to answer those questions correctly (see Davidson, 1985).

In the problem solving lesson, the teacher invited students to speculate about text content before reading. Students shared their prior knowledge about light, heat, and color and reacted to their classmates' ideas and explanations. This discussion helped students establish their own reasons for reading. Over half of the students in the problem solving lesson told us they were thinking about science concepts as they read. One student said, "I was trying to think of what to say along with what they were saying... I agreed with Robin and Don because when you read magazines it says that infrared is hot and ultraviolet is cool."

Rather than reading to answer the teacher's questions, the students in the problem solving lesson read to learn about the science concepts, to verify predictions they or others had made, and to connect text information with their own knowledge as well as with the ideas presented during the earlier discussion. These students shared responsibility for learning about light and color. The difference between the business as usual lesson and the problem solving lesson influenced the way students processed the science text. Moreover, the problem solving lesson is consistent with the scientific method used in most science classes.

Discussing the Reading

The paragraph the students read on light and color is presented in the Text Passage. After reading this portion of their texts, the students and teachers in these two lessons discussed the concepts and ideas presented.

TEXT PASSAGE

The sun's spectrum also contains light that cannot be seen. An invisible part of a white light spectrum beyond visible red light is called infrared (in fruh RED). **Infrared** has a longer wavelength than visible red light. Infrared waves are heat waves. All objects give off unseen infrared radiation. This principle is the basis of infrared photography. In general, warm objects give off more infrared radiation than cold objects. Infrared radiation passes through clouds, fog, and haze more readily than does visible light. Therefore, infrared film will often produce clearer pictures than regular film on a foggy or cloudy day.

From Heimler & Neal, 1979, Principles of Science, Book 2, 396-397, Merrill.

Business As Usual Lesson

Teacher: My first question to you is, "What is an advantage of infrared film?"

Jim: (raises hand) I think I know. It passes through fog and stuff.

Kathy: It can take clearer pictures on cloudy days.

Jim: And like transistor of clouds.

Teacher: And why does that happen? You said transistor of clouds, but what did that relate to? (pause) The infrared will pass through clouds, you said.

Jim: Fog.

Teacher: It could take better pictures. All right...

Kathy: 'Cause it's warm.

Danny: Faster than light.

Kathy: It's warmer.

Teacher: Will regular light do that?
(Several students together) No.

Teacher: OK. Read the next paragraph and don't go any farther.
(students read)

Problem Solving Lesson

Teacher: OK, were you right about anything so far?

Robin: Infrared isn't one of the hottest colors.

Teacher: It isn't? (teacher nods) Anybody else? Find out anything else that's surprising?

Mark: That infrared can go through clouds and haze faster than regular light goes.

Teacher: What surprised you about that?

Mark: I would think that infrared would stop faster than visible light does.

Teacher: OK. The second part of the paragraph talks about infrared films producing clearer pictures than regular film. What do you expect to follow in the next paragraph? It's setting you up for something. What do you think it is?

Don: They probably talk about the next hottest color. And then go on and talk about the other colors.

Teacher: All right. What do you think would be the next hottest color?

Don: Ultraviolet.

Teacher: Why?

Don: Well, I learned it. That's what I think. I'm not too sure, or anything.

Teacher: What else? Do you know anything else about ultraviolet?

Dale: That it's used for lasers. NASA has some.

Shana: Too much can be harmful.

Teacher: Too much what?

Shana: Ultraviolet light.

Teacher: How do you know that?

Shana: I just do—you read about it in magazines.

Teacher: OK.

Business As Usual Lesson		Problem Solving Lesson
	Dale:	It causes skin cancer. Like if you stay out in the sun too long, the ultraviolet rays on a clear day, it burns you. That's how you get a sunburn, and if you keep doing that day after day, you know, you stay tan all year 'round and stuff. Um, maybe after five, ten years you start developing skin cancer.
	Don:	Isn't ultraviolet the kind that comes out of fluorescent lights?
	Teacher:	Why do you think that?
	Don:	I'm not too sure myself.
	Teacher:	What makes you think it would?
	Don:	Well, it gives off that color. A sort of violet color, fluorescent lights.
	Teacher:	When we started you were not too sure what you knew about the spectrum. And you just found out that you know a whole lot of things. They're coming out, slowly but surely. Anything else?
	Steph:	Ultraviolet is harmful... the ultraviolet rays.
	Teacher:	Where have you seen this kind of light?
	Steph:	In the plant light.
	Teacher:	In the plant light. (nods)
	Shana:	They have them in hospitals, too.
	Teacher:	What would their use be there?
	Shana:	Uh, I think they're used for when they examine you, probably.
	Sam:	I've seen things in a book that ultraviolet lights can also kill germs.
	Teacher:	OK. Well, read to the end of the next paragraph. (students read)

Both of these excerpts are discussions about portions of the same textbook. Differences in the ways students and teachers interacted with one another and with the ideas from the texts are apparent. In the business as usual lesson, the teacher asked questions to see if students understood, or could remember, the main points of the selection. Students responded with factual information directly from the text. The teacher's task seemed to be one of asking questions, and the students' job was to answer them. The discussion was aimed at tying together text information.

Discussion began in the problem solving lesson when the teacher asked students whether their predictions had been confirmed. After analyzing the text information from that perspective, students speculated about the next portion of the chapter. Again, they shared their knowledge and opinions.

There was a marked difference between these two approaches. As the classes progressed, students in the problem solving lesson moved from divergent to convergent critical thinking, while the students in the business as usual lesson continued answering literal questions. The teacher encouraged students to engage in problem solving; they took more intellectual risks because the procedure freed them from fear of failure.

The patterns of interaction established early in class were evident throughout the lessons. In the business as usual lesson, students read a portion of the chapter and then answered the teacher's summary questions. In the problem solving lesson, students read, evaluated their earlier predictions, and then made new predictions.

Not only did the patterns of interaction persist throughout these two lessons, but they also characterized six other lessons we observed in social studies, literature, and science. We noted striking similarities among lessons using the same approach, regardless of teacher, students, or subject area being studied. The business as usual and problem solving lesson types represent two very different approaches to reading and discussing text, and different learning outcomes are associated with each.

Involving Students in the Lesson

We call the business as usual lesson a Teacher/Manual-Generated Lesson (TMGL), since it reflects the teacher's interpretation of the suggestions offered in the teacher's manual. TMGLs generally begin with the teacher introducing the text selection and talking with students about the concepts or issues addressed in the text. Cycles of silent reading and discussion follow.

We have analyzed the content of teacher talk and student talk during lessons using a taxonomy of inferences (see Wilkerson, 1984). The taxonomy indicates proximity to generalizations intended by the text. According to our analyses, students' talk during TMGL lessons focuses on ideas within the text. Many student comments during the lessons are simply restatements of textual material (Wilkerson, 1984, 1986, 1987). The teacher's talk also tends to make explicit reference to text information. In addition, teachers add to the discussions new information generalized from the text. In short, the teacher serves as a content expert who regulates the flow and direction of discussion and who tells students about important concepts (Padak, 1985, 1986, 1987).

Teacher regulation during the TMGLs is particularly apparent when analyzing patterns of interaction within the lessons. As noted, establishing reasons for reading is the teacher's responsibility in this type of lesson. Discussions following reading typically include summary-type questions that evoke literal responses. In addition, teachers assume responsibility for linking statements and ideas together within the lessons (Davidson, 1986). Students rarely share information or opinions with one another or elaborate on their own or others' ideas. Instead, they respond to the teacher's questions. Their attention focuses on what the teacher might say or do rather than on ideas presented in the text or suggested by another member of the group (Davidson, Padak, & Wilkerson, 1986a, 1986b, 1986c, 1987).

The TMGL framework does not facilitate integration of information from the text with students' prior knowledge and experiences. After participating in these business as usual lessons, students show little understanding of the concepts they have studied (Wilkerson, 1984, 1986).

The label we use for the problem solving type of lessons is Teacher/Student-Generated Lesson (TSGL). This label reflects shared responsibility for the learning that occurs (Davidson, 1986). The excerpts presented above came from one type of TSGL, the group Directed Reading-Thinking Activity (DRTA) (Stauffer, 1980). In the final section of the chapter, we present ideas for other TSGLs.

Analyses of student inferences during TSGLs show evidence of attempts to link prior knowledge with text concepts and content. Students explore issues related to but not resolved by the text, and they understand major generalizations from the text more successfully than do their TMGL counterparts. In short, students seem intellectually involved with the concepts and ideas (Wilkerson, 1984, 1986, 1987).

The teacher's role during a TSGL is to activate student thought, encourage use of prior knowledge, and facilitate group interaction. Purposes for reading are established through discussions involving predictions and hypotheses supported by prior knowledge and information provided in the text. In TSGLs, students determine their own reasons for reading.

Postreading discussions during TSGLs generally begin with the teacher asking an evaluative question that encourages students to confirm, modify, or reject their previous ideas with reference to the text they have just read. Student responses to these questions tend to reflect higher levels of thinking. In addition, students frequently question one another, clarify statements and text information, and elaborate on ideas during TSGLs (Davidson, Padak, & Wilkerson, 1986a, 1986b, 1987).

Ideas for Other TSGLs

Hammond's (1979) anticipation activity and Herber's (1978) reading and reasoning activities emphasize students' shared responsibility at the onset of the lesson, engage students in group discussion, and foster students' critical thinking. An example of an anticipation activity can be found in Figure 1. The teacher here asked students to categorize words from the text selection presented at the beginning of this chapter before they read the text. As indicated by the categories that resulted, students had some prior knowledge about light and color.

A reading/reasoning activity based on Herber's work can be found in Figure 2. This activity, which focuses on the properties of light, is intended for students working in small groups with a partner. Sharing responsibility for completing a task with a partner encourages student interaction and stimulates critical thinking. When the small groups finish their work, students share their understandings with the entire class. In this type of lesson, the message is clear: getting through a task isn't the goal; the thinking process that gives rise to the finished product is what is important.

Conclusions

Classroom discussions about science text are a complicated phenomenon. They are affected by social and communicative interactions among students and teachers as well as by the academic tasks at hand. In essence, students and teachers must work together to construct meaning as lessons evolve. Although the science curriculum often is described in terms of concepts, textbooks, and curriculum guides, these features do not form the whole. How teachers structure their interactions with students and how they use instructional materials define the curriculum.

Figure 1
Anticipation Activity

Steps (for the teacher)

1. Select approximately 25 key words from the reading assignment, taking care that some words come from each of the text pages.
2. Duplicate the list and cut it into sets of word cards.
3. Group students in threes and ask them to arrange the cards into categories that reflect how ideas may be related in the science text they are about to read.
4. Provide time for students to discuss their predicted categories with the class.
5. Have students read to verify their predictions.
6. After students have had an opportunity to read, suggest that they may wish to rearrange cards based on the information from the text.
7. Allow time at the end of the activity for students to discuss the changes they made or to confirm their original groupings.

One Group's Categories

spectrum
rainbow
primary colors
rays
white light
band
visible spectrum

microscope
focal length
concave lens
eyeglasses
convex lens
prisms

electromagnetic spectrum
infrared
ultraviolet
waves
energy

refracted
speed
bend
diffracted
reflected

temperature
mirage
low frequency
high frequency

Figure 2
Reading/Reasoning Activity

Part I
Directions: Work with a partner to complete the chart. You may use your textbook. Be sure you can explain your decisions.

	What is absorbed?	What is reflected?	What color will you see?
White light on a blue shirt			
Red light on a blue shirt			
Blue light on a blue shirt			
Red light on a white shirt			
White light on a white shirt			

Part II
Directions: Use the chart and the chapter to complete this section.
For each statement:

A. Tell whether it is true or false.
B. Tell why it is true or false.

1. A black object will always appear black.
2. A white object will always appear white.
3. A red object can never appear white.
4. A red object can never appear black.
5. Objects absorb all colors except their own.
6. Objects can appear white only in white light.

Students do what we ask them to do according to their interpretation of the rules of the game. If we expect them to become active participants in learning, our instructional framework must be consistent with that expectation. Teacher/Student-Generated Lessons encourage teachers and students to share responsibility for learning.

References

Davidson, J.L. (1985). What you think is going on, isn't: Eighth-grade students' introspections of discussions in science and social studies lessons. In J.A. Niles & R.V. Lalik (Eds.), *Issues in literacy: A research perspective.* Rochester, NY: National Reading Conference.

Davidson, J.L. (1986). The teacher-student generated lesson: A model for reading instruction. *Theory into Practice, 25,* 84-90.

Davidson, J.L., Padak, N.D., & Wilkerson, B.C. (1986a). *Four instructional lessons in science: The process of comprehension attainment of text.* Paper presented at the Thirty-First Annual Convention of the International Reading Association, Philadelphia, PA.

Davidson, J.L., Padak, N.D., & Wilkerson, B.C. (1986b). *Instructional frameworks in junior high science and social studies: How lessons evolve.* Paper presented at the annual meeting of the American Educational Research Association, San Francisco, CA.

Davidson, J.L., Padak, N.D., & Wilkerson, B.C. (1986c). *Instructional frameworks in junior high science and social studies: The hidden curriculum.* Paper presented at the annual meeting of the American Educational Research Association, San Francisco, CA.

Davidson, J.L., Padak, N.D., & Wilkerson, B.C. (1987). *Reconsidering a focus for curriculum development: Curricular issues.* Paper presented at the annual meeting of the American Educational Research Association, Washington, DC.

Hammond, D. (1979). *Anticipation activity.* Paper presented at the Great Lakes Regional Reading Conference, Detroit, MI.

Heimler, C.H., & Neal, C.D. (1979). *Principles of science, Book 2.* Columbus, OH: Merrill.

Herber, H.L. (1978). *Teaching reading in content areas* (2nd ed.). Englewood Cliffs, NJ: Prentice Hall.

Padak, N.D. (1985). *A teacher's verbal patterns and decision making: The influence of instructional setting.* Paper presented at the annual meeting of the American Educational Research Association, Chicago, IL.

Padak, N.D. (1986). Teachers' verbal behaviors: A window to the teaching process. In J.A. Niles & R.V. Lalik (Eds.), *Solving problems in literacy: Learners, teachers, and researchers* (pp. 185-191). Rochester, NY: National Reading Conference.

Padak, N.D. (1987). *A comparative analysis of the teacher's roles in two junior high literature lessons.* Paper presented at the annual meeting of the American Educational Research Association, Washington, DC.

Stauffer, R.G. (1980). *The language-experience approach to teaching reading* (2nd ed.). New York: Harper & Row.

Wilkerson, B.C. (1984). *A study of students' inferences during and following a group directed reading-thinking activity and a group directed reading activity in social studies.* Unpublished doctoral dissertation, Northern Illinois University, DeKalb, IL.

Wilkerson, B.C. (1986). Inferences: A window to comprehension. In J.A. Niles & R.V. Lalik (Eds.), *Solving problems in literacy: Learners, teachers, and researchers* (pp. 192-198). Rochester, NY: National Reading Conference.

Wilkerson, B.C. (1987). *A comparison of student inferencing behavior in literature lessons and student inferencing behavior in science and social studies lessons.* Paper presented at the annual meeting of the American Educational Research Association, Washington, DC.

CHAPTER 9 Science Teachers' Use of Texts: Three Case Studies

Donna E. Alvermann
Kathleen A. Hinchman

In this chapter, the reader will discover answers to questions such as: What are the assumptions that drive teachers' decisions about how actively they involve students in learning from science texts? Why do some teachers rely exclusively on textbooks, while others banish them and still others use them in moderation? Authentic classroom situations provide the context for seeing teachers arrive at the answers. The reader also will find suggestions for creating a balance between instruction that focuses on hands-on experiences and instruction focusing on information from texts.

What are the assumptions science teachers make about subject-matter texts? How do these assumptions influence teachers' and students' use of textbooks and other printed materials? How might these assumptions be changed to ensure more effective use of text materials in science classes?

This chapter explores these questions from several angles. It begins with a look at what the literature says about the assumptions that drive science teachers' decisions to use printed materials. Then case studies of three secondary school science teachers are presented to explore the influence of their assumptions on the use of text in their classrooms. Finally, suggestions are made for changing assumptions to ensure more effective use of text materials in the teaching of science.

Assumptions and the Hidden Curriculum

Educators have recommended teaching science in a manner that ties learning of the scientific process to learning of science content. However, much of reported classroom activity reflects a product or a process orientation (Sigda, 1983). Teachers' use of textbooks and other printed materials seems tied to their position at either end of this process-product continuum.

The product-oriented teacher imparts scientific information using textbooks and other printed materials as sources. In contrast, if the process-oriented teacher (who is often perceived as more effective by colleagues within the science education community) uses text at all, it is usually in the minor role of providing references and directions within the laboratory setting (Penick & Yager, 1983).

Since Jackson's (1968) *Life in Classrooms,* researchers have explored the role of both explicit and implicit, or "hidden," curricula on classroom instruction. Teachers' classroom behaviors reflect their individual interpretations of these stated and unstated curricular expectations. Students, in turn, respond to these expectations as they perceive them. Hidden curricula can send students unintended messages. For instance, a science teacher who believes that a process emphasis is important in instruction may know that students are responsible for passing a product-oriented final examination. Consequently, this teacher may *attempt* to reflect both orientations while teaching, but may signal clear expectations regarding only the latter. On the other hand, a process-oriented teacher who totally disregards the textbook implicitly tells students that reading is not important in scientific pursuits.

Recent research supports the idea that teachers' behavior may send students implicit messages regarding reading. Smith and Feathers (1983) found that even when teachers make clear that the text is the primary source of information, students identify the teacher as the primary source and find ways to gain needed information without reading. Ratekin and his collegues (1985) discovered that teachers ask students to use text to review information covered in lectures. Their research suggests that students may perceive text as serving only a "safety net" purpose. Davey's (1987) survey of secondary school teachers indicates that they give students little or no instruction in how to learn from assigned readings, yet still expect them to learn in this manner.

Other research has found that textbooks serve many different purposes within the confines of teacher-directed discussions at the secondary school level (Alvermann, 1987; Alvermann et al., 1985). These purposes include refocusing discussion and other nontext-related functions. Thus, although teachers may think of textbooks as important information sources (Vacca & Vacca, 1989), students may think otherwise. Confusion reigns in the minds of learners when they receive mixed messages regarding the importance of texts and their function within and across subject matter classes.

The Case Studies

Data for the following case studies were collected from larger, long term studies that explored teachers' use of text in classrooms (Alvermann et al., 1985; Hinchman, 1987). Both of these larger studies used qualitative research methodologies (Bogdan & Biklen, 1982; Glaser & Strauss, 1967) that involved collecting and analyzing data in schools over the course of a year.

Each of the teachers in the case studies was identified by colleagues as an effective teacher. A variety of classes were observed, teachers were interviewed, and relevant classroom documents were collected. The resulting field notes, videotapes, interview transcripts, and classroom documents provided information for this chapter.

Teacher A

Teacher A had a delightful sense of humor and was liked and respected by her colleagues. She taught ninth grade general science and tenth grade college preparatory biology at a large metropolitan high school. Previously, she had been a substitute teacher and a middle school science teacher. She was an organized person who methodically balanced home and school obligations.

Teacher A had a product orientation. She felt that the most important part of her job was tied to the population she was hired to teach.

She wanted to help innercity students who did not necessarily see themselves as college material. For her ninth graders, she believed she could best accomplish this goal by preparing them to pass the district final exam. Teacher A used this test and the textbook as the basis for her planning. Her organization followed a contract system that she felt provided a means of managing student behavior.

At the start of a unit, students were presented with a list of activities they were to accomplish before they were allowed to take a unit test. Some were laboratory activities with printed directions, but most involved answering literal-level text questions. Much class time was spent with students sitting in their seats matching key words in questions to information in their texts. Their work was collected and graded on a regular basis. A typical class period would find students approaching the teacher's desk in the front of the room with paper and book in hand to ask questions about assignments.

Mary [indicating a section of text]: I don't get this one.

Teacher [flipping through the text and pointing to a heading]: If the book says this one is 2,000 times larger, and the question asks which is bigger, what are you going to write, Mary?

Mary [pointing to a picture]: This one.

Teacher Good. [Student returns to her seat]

Sometimes this cycle was repeated several times before the student returned to his or her seat. When several students asked the same question, the teacher would offer an explanation to the whole class. Judging from their actions, students appeared to see the teacher's interpretation of the text as a source of information in the class.

Teacher A used the textbook and her knowledge of the statewide final exam to structure her tenth grade biology class. This class followed a lecture format, with students copying notes from an overhead transparency as the teacher spoke.

Teacher: We're at your favorite word, *phagocytosis.* Amoebas are fresh water animals. You can find them in the lakes and ponds around here. So every time you go swimming and take a big gulp of water, think of all the amoebas you're eating.

Students: Yuck.

John: Do we have much more to write?

Teacher: A bit.

Susan: Can you get sick from eating them?

The lecture continued for the entire period, with occasional questions from students. References to the text were included in the teacher's notes. At the end of a section of notes, the teacher assigned the questions from the end of the text chapter. These questions seemed to be used to reinforce the lecture, which was the primary source of information. After students had submitted their completed assignments, papers were graded and returned, so again, students were encouraged to see the text through the teacher's interpretation. Teacher A reviewed students' answers through whole class discussion. Occasionally, the students used the textbook as a reference tool when they completed laboratory observations.

Teacher A said that both of her textbooks were selected by district committees whose concerns focused on the fit between the required curriculum and students' general reading ability. She used these books in both classes, although she noted the need to adjust their use because the reading level was too high for some of her students. Literal understanding of Teacher A's text interpretations was further reinforced in both classes by end-of-unit tests. The teacher constructed the tests to reflect the multiple-choice format of the district and statewide final exams. Reviews before tests consisted of question and answer periods that reflected the factual nature of these tests.

Teacher A's instruction and her use of texts seemed tied to the product orientation of the final exam the students needed to pass to be successful in her classes. The teacher and her students referred often to what "they" (state education officials) wanted as a means of deciding how to answer questions based on the text. Teacher A explained: "The state gives me an outline: We want you to cover x, y, and z. And it's up to my discretion and my organization as to how I will present it. But I've got to know in my head that by the end of May I have to make sure that I've covered a, b, c, and x, y, and z, and pad it with things of interest to the kids. Diversity.... What I am interested in and what we have time to talk about."

Teacher B

Teacher B had a master's degree in biology and was enrolled in a specialist degree program at a nearby university during the time we observed his classroom. As chair of the school's science department, Teacher B was respected by his colleagues for his integrity and by his students for the personal interest he took in each one of them. Teacher B had grown up in the same rural, economically depressed area in which he taught. His well-stocked aquarium, which was the focal point for many of his process-oriented lessons, was evidence that he knew how to motivate his eighth grade general science students.

Teacher B was an avid reader. During his daily planning period, he could be found in the school library, deeply engrossed in the latest issue of *Scientific American* or some other science periodical. To say that the students in his classroom were unenthusiastic readers would be an understatement. They read as little as possible, although reading was not a problem for most of them. Teacher B had all but retreated from using the textbook in his general science classes because he believed the students would find science boring if they had to read their texts. From the following transcript of one of Teacher B's general science lessons, it is clear that the text played a minor role in students' learning. They were expected to use their eyes for observing, not for reading. In short, process, not content, was important.

Teacher: Think about the fish tank here in the room. What are the members of this community?

Students [calling out answers one at a time]: Algae.
Goldfish.
Suckerfish.

Teacher [writing on board]: In a community, animals live in a specific area—what do we call the area?

Students: Habitat.

Teacher [writing *habitat* on the board]: The book also gave a name to a special place in the habitat.

Donald: Niche.

Teacher: Is it possible, Donald, for two organisms to have the same habitat but different niches?

Donald: Yes.

Teacher: Which fish in the tank have different niches—different jobs? Donald?

Donald: Goldfish.

Teacher: Okay. What is the occupation of the goldfish?

Donald: It swims back and forth.

Teacher: How about the suckerfish?

Donald: It stays on the bottom—eats stuff.

Teacher: Veneeta, which two fish were having a dispute last week?

Veneeta: The crayfish was in the suckerfish's territory, near the bottom.

Teacher: Now, can you see why there was some conflict?

The lesson continued for approximately 40 minutes, but only one additional, brief reference was made to the text despite the fact that the teacher had assigned part of the chapter on ecosystems as homework. Students could participate in the class discussion without having read their assignments, and because the chapter test was based on class discussion and class notes, it was possible to ignore the text altogether. Students in this class interpreted the hidden curriculum correctly: reading was not important to learning science concepts.

Teacher B was not bothered by the mixed message he gave his students when he assigned material to be read from the textbook and then ignored the text during class discussion. He said he had decided against shelving the textbook completely, as the other general science teacher had done earlier in the year. He did not give a reason for his decision.

This teacher saw little hope for reading as a tool students could use to learn science content. This attitude, however, did not mesh with his own behavior—his almost daily visits to the library to read popular science journals. In an interview with one of the researchers, Teacher B explained his seemingly incongruent behavior this way: "I am reading to enrich my knowledge of science; I already have the basics." He also said that he did not think he knew enough about reading to help students who had difficulty with the textbook. Judging from the feedback he received from students, Teacher B assumed they enjoyed a process approach to science, with its emphasis on observing, predicting, experimenting, and interpreting.

Teacher C

Teacher C had more than 9 years of classroom experience as a middle school science teacher. She team-taught with a group of individuals who were active professionally and who in the past had collaborated on several projects with teacher educators from a nearby university. Five years earlier, she cooperated in a doctoral student's dissertation study involving wait-time training. At the time of our observations, vestiges of that training were still evident. Unlike several other teachers we observed, Teacher C gave students time to reflect on her questions before moving on.

Teacher C differed from the other teachers in another important way. She expected her students to learn from their science textbooks, but the text was only one of several information sources available to them. Students conducted experiments, studied visuals and diagrams, observed and recorded physical phenomena using charts, and checked their predictions against the data they collected as well as against the text and their own knowledge of how things work in the real world.

The students in Teacher C's science class were average to below average in reading ability. Most of them found the science text difficult but not impossible to comprehend. Part of their success in comprehending the concepts put forth in the text could be attributed to Teacher C's knowledge of how important students' past experiences were to their current learning. Consider, for example, how she introduced students to the difference between rolling and sliding friction by involving them in an experiment in which they could use their prior knowledge to help solve the problem.

Teacher [rolling her chair out from behind her desk]: Dan, what distinguishes my chair from the one you're sitting on?

Dan: Mine doesn't have wheels, but everything else looks the same—no, the color—that's different.

Teacher: Okay, mine has rollers, or wheels, and your chair doesn't. Suppose I asked Dan to come up here with his chair and I had him push first his chair and then my chair. Which chair do you predict would go farther, and why? Before we try this little experiment, take out a piece of paper and write your predictions.

Dan: Do you want me to bring my chair up now?

Teacher: Yes, but write your prediction first, and all of you should be sure to state why you think one chair will move farther than another. Or maybe you think they'll go the same distance; that's okay too. Just tell why you think so.

Teacher [after giving students time to make their predictions]: Okay, Dan, ready?

Dan [pushing both chairs with what appears to be an equal force]: I was right! I predicted your chair would go the farthest.

Teacher: Chris, come up and measure the difference, please. What about the rest of your predictions? Were you right or do you need to change them?

Most of the students said they had predicted the same thing as Dan. However, in a brief follow-up discussion, it was clear the students did not agree on why the teacher's chair moved farther. One student suggested the experiment should be run again, this time with a girl pushing the two chairs. The teacher agreed, and the procedure was repeated. In fact, the experiment was repeated several times, and students recorded their predictions on a chart that included information such as the experiment number, the difference in distance traveled, and the explanation for the difference. The discussion that follows is an indication of what the students knew before reading the section in their textbook on the different types of friction.

Vanessa: It has something to do with motion, I know that much. Maybe that law we learned about last week? What was...

Dedrick [interrupting]: Newton's law?

Vanessa: Yeah, that something stays still or keeps moving until a force changes it.

Teacher: Okay, what was that called?

Several: The law of motion.
Inertia.
Gravity.

Teacher [interrupting]: Okay, it was inertia, wasn't it? Inertia—okay, we know that it took a force like Dan to start the two chairs moving, but what was the force that made the chairs stop moving?

In the brief discussion that followed, none of the students explained the phenomenon they had observed in terms of rolling versus sliding friction. Consequently, the teacher suggested they turn to the chapter on motion, which described how sliding friction produces a greater force than rolling friction. She also called their attention to the illustration that compared low rolling friction (a train wheel on a track) with high rolling friction (a bike tire on the pavement).

Because the students had a purpose for reading that grew out of a problem they could not solve using their own knowledge, they were motivated to read the textbook explanation. Teacher C did not stop with that explanation, however. The students' assignment for that day was to think of an answer to the teacher-prepared worksheet puzzle:

> If the two chairs in the classroom experiment were placed aboard a spaceship bound for the moon and each was pushed with the same amount of force, would they stop? Would one go farther than the other? Why?

Analysis of the Case Studies

The three science teachers used their textbooks in quite different ways. Teacher A's use of text reflected her methodical, product-oriented approach to covering the curriculum. Her students were expected to glean explicit facts from the text to complete their assignments. The literal-level multiple-choice questions on Teacher A's tests reinforced the manner in which she used the textbook. Reviews before tests further emphasized the importance she attributed to the text. It is likely, however, that the students had become so dependent on Teacher A's explanations that they had all but given up trying to interpret the text for themselves.

Teacher A's stated instructional goals were admirable given the setting and curriculum constraints within which she taught. Nevertheless, if she were to follow Sigda's (1983) recommendation and fuse process and product orientations to form instructional goals, students might benefit from improved long term understanding of science content and processes without any loss in test performance. In addition, fusing process and product orientations would eliminate the need to cover the same information twice, once in lecture and once with the text.

Teacher A might also consider varying how she uses the textbook so that sometimes she would be the primary source of information, sometimes the text would be the primary source, and sometimes hands-on laboratory activities would allow students to be the primary source. Teacher A would need to keep in mind that when the text is the primary source of information, students should be encouraged to work through their own interpretations of the content. This goal might be accomplished by having students first complete study guides and then hold small group discussions. Suggestions of other activities can be found in professional journals and in various content area methods texts (Herber, 1978; Readence, Bean, & Baldwin, 1989; Vacca & Vaccca, 1989).

Teacher B's nonuse of the text reflected his belief that students would refuse to accept any type of science instruction other than process-oriented instruction. Having grown up in the same geographic area in which he was teaching, Teacher B assumed he knew what would motivate students to learn. In truth, the aquarium was a big success, and Teacher B found many creative ways to use it. Like Teacher A's plan, however, Teacher B's curriculum focus lacked balance, which resulted in less than optimal instruction. Instead of having their attention focused on the text (as was the case for Teacher A's students), students in Teacher B's room were forced to depend on their own knowledge for generating new ideas. Although textbook assignments were routine, they were rarely discussed. Students got the impression that anything worth learning about in science could be acquired through first-hand experiences rather than through books or other printed matter.

Because Teacher B was an avid reader of science material, he would make an excellent role model for students. If he were to bring popular science magazines to class or encourage students to find library books that dealt with some of the same topics the class was studying, the course content would be enriched and a better balance between process and product would be possible.

Teacher B said he was reluctant to use the text because he didn't know enough about reading to help students when they had difficulty comprehending. As a science teacher, however, he possesses knowledge as valuable as that of any reading teacher. He knows how scientists categorize information. He knows that when scientists write about the habitat of certain species, they may categorize information according to parts, properties, location, and function. By cueing students into this categorizing system and others like it (see both Armbruster and Harrison, this volume), Teacher B would improve his students' comprehension and retention of science text without detracting from the process approach he so firmly believes in.

Teacher C's use of the text reflected a balance between process and product. She involved students in applying their science and reading process skills and at the same time had them read for information. Teacher C expected students to learn from text, and she made certain they had the motivation and skills to meet her expectations.

Students had fewer occasions to become confused over the role of the textbook and other printed materials in Teacher C's room. Questions that arose during a science experiment on friction eventually were answered by reading the text. Conversely, a question that came up later during a discussion of the text gave rise to a science activity. This balance between process and product conveyed to students that sometimes questions and problems are better answered by experimenting and sometimes they are better answered by reading. No hidden curriculum drove the use of printed materials in Teacher C's room—or, if it did, it was not as evident as in the other two classes.

Conclusions

Describing science teachers' approaches along simple process-product lines does not account for the complex real life variables involved in teachers' choices of instructional practices. However, we believe this strategy has value as a tool for examining some of those variables, including the assumptions that underlie many textbook practices.

Instruction can be designed to help students learn science content at the same time they are learning science and reading process skills. Such instruction will not depend on text as the sole information source or on the elimination of text in favor of hands-on activities. Rather, effective instruction will draw from several information sources, including a variety of texts as well as the students' own experiences. This type of instruction will enable students to gain more sophisticated understandings of science and will do much to alleviate the confusion caused by the hidden curriculum.

References

Alvermann, D.E. (1987). The role of textbooks in teachers' interactive decision making. *Reading Research and Instruction, 26,* 115-127.

Alvermann, D.E., Dillon, D.R., O'Brien, D.G., & Smith, L.C. (1985). The role of the textbook in discussion. *Journal of Reading, 29,* 50-57.

Bogdan, R., & Biklen, S. (1982). *Introduction to qualitative research in education.* New York: Aldine.

Davey, B. (1987). How do classroom teachers use their textbooks? *Journal of Reading, 31,* 340-345.

Glaser, B.G., & Strauss, A.L. (1967). *The discovery of grounded theory: Strategies for qualitative research.* New York: Aldine.

Herber, H.L. (1978). *Teaching reading in content areas* (2nd ed.). Englewood Cliffs, NJ: Prentice Hall.

Hinchman, K. (1987). The meaning of the textbook for three content area teachers. *Reading Research and Instruction, 26,* 247-263.

Jackson, P. (1968). *Life in classrooms.* New York: Holt, Rinehart & Winston.

Penick, J.E., & Yager, R.E. (1983). The search for excellence in science education. *Phi Delta Kappan, 64,* 621-623.

Ratekin, N., Simpson, M.L., Alvermann, D.E., & Dishner, E.K. (1985). Why teachers resist content reading instruction. *Journal of Reading, 28,* 432-437.

Readence, J.E., Bean, T.W., & Baldwin, R.S. (1989). *Content area reading: An integrated approach* (3rd ed.). Dubuque, IA: Kendall/Hunt.

Sigda, R.B. (1983). The crisis in science education and the realities of science teaching in the classroom. *Phi Delta Kappan, 64,* 624-627.

Smith, F.R., & Feathers, K.M. (1983). The role of reading in content classrooms: Assumption vs. reality. *Journal of Reading, 27,* 262-267.

Vacca, R., & Vacca, J. (1989). *Content area reading* (3rd ed.). Glenview, IL: Scott, Forseman.

PART 4

Practical Applications

Ideas for teaching fill the chapters in Part Four. In each case the authors walk us through instructional strategies applicable to practically any grade level. They include strategies for a variety of activities ranging from teaching reading for main ideas and procedures for organizing important information to instruction in vocabulary, writing, and imagery.

Aulls leads off with a chapter on main ideas. It makes sense to begin here. Reading for main ideas is critical for learning content, as well as for succeeding with other learning strategies (for instance, notetaking, summarizing, self-questioning). The main idea theme is carried forward in the next two chapters on framing. Frames are visual representations of important content that depict main ideas and their interrelationships in chart format.

Although the three remaining chapters focus on diverse topics, the authors share the same goals. In each instance they stress active student involvement, teacher modeling, and the gradual shift of responsibility for learning from teacher to student. Ideas flow from these chapters. Most are not specific to any one grade level; you are sure to find several adaptable to your classroom.

CHAPTER 10 *Main Ideas: Key to Learning Science*

Mark W. Aulls

This chapter begins with a brief discussion about our goals as teachers. Do we want our students to be forever dependent on us as teachers, or should we teach for student independence? While the answer may be obvious, teaching for student independence is not a simple task—particularly when the goal is to identify important and often difficult concepts in science texts. Aulls models a strategy that helps middle grade students become more independent in comprehending important science concepts. He then offers practical suggestions for adapting the approach within any classroom.

Science instruction in grades 4 through 8 entails teaching students to think like scientists, to solve problems, and to comprehend the content presented in natural science materials, including science textbooks. However, students often have difficulty reading their science texts. Students in these middle grades find expository prose more difficult to understand than narration (Aulls, 1975; Calfee & Curley, 1984; McGee, 1982; Taylor & Samuels, 1983). In particular, students have problems identifying main ideas and relationships among key topics in science textbooks.

The difficulty rests with both the texts and the students. Main ideas are often obscured by extraneous information (Holliday, this volume) or only implied within poorly developed paragraphs. Students must know how to sift through the material to arrive at the central message in a selection.

It is not surprising that students learn more when main ideas and topics are clearly presented in the materials. Baumann (1986) verified this commonsense notion in a study in which he rewrote science passages from four popular textbooks. In these revisions, he presented general topics in the titles and subheadings and made sure main ideas were explicit in the paragraphs. Fifth grade students read either the original passages or the revisions. Then they took a test that asked them to write the main idea of the entire passage, write the main idea of each paragraph, and select statements representing main ideas in the passage. On most of the three comprehension tasks, students reading the rewritten passage outperformed those who read the original passage.

Clearly, publishers need to produce science materials designed to signal important information. Although progress has already been made in writing more considerate texts (Meyer, this volume), main ideas should be more explicit in future texts.

Even if texts become more considerate, students will benefit from instruction in how to identify important information in science text. This chapter provides a rationale for teaching main ideas as part of science and offers evidence for the feasibility of this endeavor.

Three Study Reading Approaches

The competent reader constructs a network of important information rather than unconnected ideas and unimportant details (Calfee & Curley, 1984; Van Dijk & Kintsch, 1983). Teachers use a number of procedures to help their students understand main ideas from their textbooks. At the risk of oversimplifying, they can be divided into three approaches.

Approach One

With Approach One, the teacher develops prereading aids and presents them to the class. These might be graphic organizers (Alvermann, 1981, 1982; Alvermann, Boothby, & Wolfe, 1984), structured overviews (Herber, 1978), or semantic maps (Armbruster, 1979). Research suggests these study aids work. With each of these methods the teacher reveals the semantic array of topics in the selection and their interrelationships. When students have a view of these interrelationships before reading their assignments, they remember and understand more information, even with science textbooks (Hawk, 1986). Through visual representations and discussions, the teacher shows the students how to process important information.

This approach is a short term solution because students' success in internalizing the content depends solely on the teacher's knowledge of the important ideas in the text. Students do not learn about cues that signal important information in exposition or how to extract that information on their own. Moreover, these strategies place students in a passive learning role, since the teacher constructs the representations.

On the positive side, Approach One provides a means of signaling important concepts and their interrelationships before students read. As a result, students can form a mental map that guides their reading of the material and makes the ideas presented easier to recognize and understand.

Approach Two

A second instructional approach is to teach a global study reading strategy or tactic. This approach differs from Approach One because the teacher places more stress on teaching students how to use the reading strategy on their own. Global strategies include one or more forms of procedural knowledge, or knowledge of how readers construct relationships among parts of text. An example is knowledge of the steps one goes through to take notes or write a summary of a selection. With this approach, the teacher focuses on helping students analyze and learn the procedures to succeed with a study reading task.

Global strategies that have been found to be both teachable and beneficial for reading comprehension include hierarchical summarizing (Taylor & Beach, 1984), networking (Dansereau, 1985), and SQ3R (Adams, Carnine, & Gersten, 1982). One problem with teaching global strategies is that students who show immediate gains in using these strategies often stop using them on their own or use them with decreasing success after the initial instruction (Adams, Carnine & Gersten). Thus, instruction in procedural knowledge alone may not lead students to internalize the strategy or regulate their use of it.

Yet global strategies are ahead of those described in Approach One because students acquire (and at least partially internalize) knowledge about procedures for reading and studying and thus become less dependent on the teacher. Moreover, students who learn and use such a strategy take on a more active mental role as they process different pieces of textual information. With Approach Two, however, the focus is on teaching the procedures of certain strategies and not on helping students apply the strategies to text structure. This focus keeps students somewhat dependent on the teacher.

Approach Three

The third instructional approach extends the learning goals for teaching a global reading strategy. With this approach, students fully internalize the use of a strategy. Teachers help students accomplish this goal by stressing procedural knowledge as well as text knowledge. Students do more than just learn the series of steps involved in summarizing, taking notes, or asking questions; they also learn which text features signal important statements. Students who combine their knowledge of text structure with a global reading strategy should be more successful and independent.

Knowledge about how text is organized and written allows readers to evaluate their own notes, summaries, and questions by comparing them with the main ideas in the text. Such dual knowledge helps students monitor their own learning. In sum, students make use of their metacognitive skills (Baker, this volume) and use strategies more spontaneously across a variety of learning situations. Therefore, the third approach represents a long term solution to improving reading and studying.

Improving Sixth Graders' Comprehension of Life Science

A small but growing body of research suggests that the global study strategies of Approach Three produce a lasting influence on students' learning (Michaud, 1988; Wong, 1986, Wong & Jones, 1982). Recently, I conducted a study that lends additional backing to this body of research. The results of the study provide strong support for the type of instruction advocated in Approach Three. Because a full description of the study has been published elsewhere (Ritchie, 1985), our goal is to highlight only the instructional components of the method used.

For 6 weeks, the researchers had a sixth grade teacher spend half (30 minutes) of each daily reading period teaching students how to identify and infer general topics and main ideas from science selections. The teacher used an additional 3 weeks to teach students a reading strategy in which they paused after each paragraph to identify or generate a main idea or general topic and then asked a question about it. Given the importance and difficulty of the task, 9 consecutive weeks of instruction seemed to be the minimum time needed to allow students to internalize the strategy.

Three instructional procedures were evaluated. Group A students learned how to identify and infer general topics and main ideas. The teacher modeled how to use main ideas and details for developing questions and how to evaluate questions by referring back to the text. For Group B, the teacher taught topics and main ideas but did not show students how to analyze text to develop and evaluate their own questions. In Group C (the control group) students received no specific instruction in either identifying main ideas or generating questions.

The researchers predicted that students in Groups A and B would better comprehend important science content than those in the control situation. Moreover, students in Group A were predicted to perform better than those in Group B. Learning how to write questions based on main ideas and topics should reinforce main idea identification. In addition, teaching students how to identify stated main ideas and general topics should be reflected in the quality of their science questions. Thus, the researchers expected Group A students to write better questions—reflecting main ideas in the science text—than students in the other two groups. The data provided evidence that these predictions were correct.

The study extended over 12 weeks. For 9 weeks, the teacher spent 30 minutes a day on instruction. Weeks 1, 7, and 12 were used for testing. During the first week, all students took a 48-item main idea test (Aulls, 1975), which revealed no significant differences among groups. In general, scores were low, which confirmed that the students lacked skill in identifying main ideas. During week 12, students read a 750-word excerpt from a life science text and completed a literal comprehension and main idea test. Students also were asked to generate questions about the main idea of each paragraph.

Group A students did far better than the others on the posttest assessing main idea comprehension—they got 98 percent of the items correct, compared with 73 percent for Group B and 55 percent for Group C. The main idea instruction covered during the previous 6 weeks, along with instruction on how to ask topic and main idea questions, definitely made the greatest difference in student performance. Moreover, there were differences in the quality of questions the groups wrote on their own while studying science materials. Group A students generated more questions focusing on main ideas and topics than did students in Group B. The questions generated by Group B students focused more on unimportant details; seemingly, these students did not understand how to analyze text to develop questions focusing on essential information.

Given the success students had in reading for main ideas and self-questioning, it seems worthwhile to take a closer look at the instructional routines for Group A. The students received explicit instruction in main idea skills for 6 weeks, followed by self-questioning instruction for 3 weeks. Their instruction booklet included five instructional units. Each unit contained an explanatory, a guided practice, and an independent phase followed by self-evaluation and a teacher conference.

The teacher worked with the students through the explanatory phase of each unit. He explained why analyzing text and understanding procedures for focusing on key points helped in study reading. For the first three lessons, he taught students how to categorize and classify words, phrases, and sets of sentences into superordinate and subordinate relationships. Students learned to cluster words and then sentences into categorical relationships.

In later units, students learned what a main idea is and how to distinguish between the general topic and the main idea in a paragraph. The teacher stressed why understanding this distinction is critical for coming up with main ideas, particularly when main points are hidden within the text. Once they understood this difference, they were ready to learn step-by-step procedures for identifying topics and main ideas.

Students also learned how to identify cues for locating main ideas within paragraphs. The teacher modeled how to derive the main ideas and topics from the text. He also taught them when to reread paragraphs to confirm their predictions about the statement they believe to be the main idea. A description of these instructional procedures appears in Figure 1.

In the last 3 weeks of the study, the teacher taught students how to generate a question for each paragraph as they read selections from their science textbook. Each passage came from a life science textbook and ranged from 500 to 2,500 words. The selections were not altered in any way from the materials used for science instruction in grade six classrooms.

Each lesson required one to two 30-minute reading periods to complete. The teacher began by explaining the strategy of self-questioning during reading and why it helps comprehension of science selections. Then he modeled aloud the thinking steps to use in generating questions focusing on important information. He took turns with the students thinking out loud, demonstrating how to generate main ideas and a question for each paragraph.

After modeling, he asked students to try this process by thinking aloud while they ana-

Figure 1
Teacher Think Alouds

Begin instruction by modeling for students the thinking processes you use when reading for main ideas. This involves selecting, rejecting, and paraphrasing in order to arrive at the central point of a paragraph or a selection. It is helpful to teach students the difference between topics and main ideas. A topic is a one or two-word description that tells what the paragraph or section is about. Often it is printed in bold print as a subhead, or it might be repeated as the subject of one or more sentences in the selection. The main idea is a sentence containing the most important information about a topic.

Sample Think Aloud

Read aloud a section of your science text. Talk about how you determine main points and details as you read. Consider using the following steps:

Step 1 First I need to determine what the entire selection or paragraph is about. What is the topic?

- Are there bold print topics and key vocabulary that can help me with this section?
- Are any ideas repeated?

Step 2 Now that I know the topic of the selection, I am ready to determine the most important information about the topic. I am going to read the first paragraph and see if the author presents the main idea in one of the sentences.

Step 3 If there is no overall statement for the main idea, I need to come up with my own main idea for the selection. How do I do this?

- The first thing I do is come up with a topic of the paragraph or section.
- Next, I read through the sentences in the paragraph and make a list of all of the ideas or supporting details that tell about the topic.
- Then I think about the details and the topic and come up with the main idea.

Evolving Procedural Steps

After reading and thinking aloud, lead a discussion about the steps you used to come up with the main points. Write the steps on a chart or some other permanent place for easy reference so that students can begin internalizing the process. The steps might look something like this:

Step 1 Come up with the topic.

Step 2 Search for the main idea sentence in the paragraph. If there is no main idea statement, go to Step 3.

Step 3 Read through the sentences in the paragraph.

Step 4 List details that tell about the topic.

Step 5 Think of a main idea that tells about the topic and the details.

Figure 2
Experimental Teaching Procedure

Today we are going to use the rules for finding the main idea to learn how to study more effectively. Using this method regularly when you study will help you understand and remember what you read.

The first step in this method is one you already know. Use the rules I have taught you to find the main idea. In the second step, you will make up a question whose answer will contain the general topic, the main idea, and as many subtopics as possible. Listen while I use the method and do my thinking out loud. Here is a paragraph about fish; read it silently while I read it aloud.

> Fish need oxygen to give them energy, and they get it from the surrounding water. As they swim along, water enters the mouth and passes over the gills. The oxygen vessels are in the gills. In animals that breathe air, such as birds, snakes, horses, and humans, the blood vessels in the lungs take in oxygen from the air (Ritchie, 1985).

The first question I will ask myself is, What is this paragraph about? As I read, it's clear to me that the paragraph is about fish, so I won't have to use any helping questions to find the general topic.

The second question I ask myself is, What is the author trying to tell us about fish? If I look to each sentence and try to find the general ideas, I should be able to answer that question. The second and third sentences are less general than the first and tell specifically how fish breathe. The fourth sentence is just as general as the first sentence but is not about fish and, therefore, can't contain the main idea. Thus, the first sentence is the main idea. The author is trying to tell us that fish need oxygen and that they get it from the surrounding water.

The third step is to write a question whose answer will contain the main idea, general topic, and as many subtopics as possible. A good question would be: "How do fish get oxygen from the water?" Or "How do fish breathe?" The answer to both questions would contain all of the elements we are looking for.

It's your turn now. I want you to use the rules I have taught you with the second paragraph and generate a question for it. Be ready to back up your answer by telling us the rules you used to get the answer.

---

lyzed the next paragraphs. The teacher used criteria for topic and main idea selection that students had learned in the first 6 weeks in order to show them how to confirm whether the answer to their question was indeed the main idea of the paragraph. After two lessons, the teacher gave progressively fewer verbal think alouds and encouraged students to model for one another. An example of this teaching procedure appears in Figure 2.

Throughout the nine lessons the teacher gave students feedback on the processes they used to identify main ideas and to generate questions. He also encouraged them to evaluate one another's questions. When giving feedback, the teacher reinforced the extent to which the students used the steps he had modeled and explained. He encouraged students to confirm whether the answers to their questions satisfied the criteria learned for identifying or confirming a main idea or topic.

The experimental conditions for Group A relate to the earlier discussion about the three instructional approaches. The experimental teaching situation fits best into the third approach. Students became independent in self-questioning about main ideas because they learned how to tease apart the text to arrive at main points and ask good questions. They had the skills to forge ahead on their own.

Implications for Classroom Instruction

The results of this study offer several important implications for the classroom teacher. First, reading for main ideas lays the foundation for learning content. Highlighting important information sets the stage for all learning strategies. Notetaking, self-questioning, and summarizing all hinge on being able to identify main ideas.

Second, successful teaching depends on clearly defined definitions and procedures. Students must recognize topics and main ideas and how they differ. Once you have developed these definitions, write them on a chart, clearly visible to the students. A class chart might look like the one in Figure 3.

Figure 3
Class Chart

Topic:	A one- or two-word description that tells what the paragraph is about.
Main Idea:	The most important information about the topic.
Subtopics:	Ideas about the topic.
Question:	The question form of the topic and main idea.

Refer to the chart often as you model your own thinking and encourage students to use the rules. Having this information available helps students determine whether their own topics, main ideas, and questions match the definitions.

Third, main idea instruction should follow a progression from simple to more complex. Once you have developed the rules, begin teaching with selections containing obvious topics and main ideas, such as the following paragraph:

> Plants use different methods to protect themselves from people and animals. Some plants, such as roses, have thorns that cut intruders who might trample them. Other plants, like nettles, have hairs on their leaves that sting when they are touched. Still other plants, such as poison ivy, protect themselves by secreting a chemical that is dangerous to touch (Aulls & Holt, 1988).

Using the rules shows students how to answer questions from the class chart (see Figure 4).

Once students have some success with well-structured paragraphs, model how to derive topics and main ideas from selections containing implied main ideas.

The fourth implication is that teacher modeling and guided practice are absolutely essential. Telling students is never enough; demonstration is critical. Initially, doing your own modeling may be a bit intimidating, particularly if you are struggling with a difficult piece of material that has no obvious topics or main points. This struggle is essential for students to see. Talk aloud as you work your way through the material, and let students hear your thinking. Students need to know how you go back and forth and sort through information until you arrive at a topic, main idea, and question. Students also need to know that

Figure 4
Completed Class Chart

What is the topic? *Plants.*

What is the main idea? *Plants use different methods to protect themselves.*

What are the subtopics? *Roses have thorns, nettles have hairs, and poison ivy has chemicals.*

What is the question? *How do plants protect themselves?*

these skills don't simply emerge after a quick reading, and that the struggle is enticing and challenging.

Once you have modeled these strategies, have students take turns being the teacher. Let students read a selection and talk about how they came up with topics and main ideas. Also consider using cooperative reading groups. Working in groups of three or four, students can take turns reading and modeling, while the others provide feedback. Remember, students need many opportunities for supportive practice to become independent learners.

References

Adams, A., Carnine, D., & Gersten, R. (1982). Instructional strategies for studying content area texts in the intermediate grades. *Reading Research Quarterly, 18,* 27-55.

Alvermann, D.E. (1981). The compensatory effect of graphic organizers on descriptive text. *Journal of Educational Research, 75,* 44-48.

Alvermann, D.E. (1982). Restructuring text facilitates written recall of main ideas. *Journal of Reading, 25,* 754-758.

Alvermann, D.E., Boothby, P.R., & Wolfe, J. (1984). The effects of graphic organizer instruction on fourth graders' comprehension of social studies text. *Journal of Social Studies Research, 8,* 13-21.

Armbruster, B. (1979). *An investigation of the effectiveness of mapping as a study strategy of middle schools.* Unpublished doctoral dissertation, University of Illinois at Urbana-Champaign.

Aulls, M.W. (1975). Expository paragraph properties that influence literal recall. *Journal of Reading Behavior, 7,* 391-400.

Aulls, M.W., & Holt, W. (1988). *Active composing and thinking* (Act II). Dubuque, IA: Kendall/Hunt.

Baumann, J.F. (1986). Effect of rewritten textbook passages on middle grade students' comprehension of main ideas: Making the inconsiderate considerate. *Journal of Reading Behavior, 18,* 1-22.

Calfee, R.C., & Curley, R.G. (1984). Structures of prose in the content areas. In J. Flood (Ed.), *Understanding reading comprehension* (pp. 161-181). Newark, DE: International Reading Association.

Dansereau, D.F. (1985). Learning strategy research. In J.W. Segal, S.F. Chipman, & R. Glaser (Eds.), *Thinking and learning skills* (Vol. 1, pp. 209-240). Hillsdale, NJ: Erlbaum.

Hawk, P.P. (1986). Using graphic organizers to increase achievement in middle school life science. *Science Education, 70*(1), 79-87.

Herber, H.L. (1978). *Teaching reading in content areas* (2nd ed.). Englewood Cliffs, NJ: Prentice Hall.

McGee, L.M. (1982). Awareness of text structure: Effects on children's recall of expository text. *Reading Research Quarterly, 17,* 581-590.

Michaud, D. (1988). *The differential influence of knowledge of signals to importance on eighth graders' accuracy in representing content and organization of essays.* Unpublished master's thesis, McGill University, Montreal, Canada.

Ritchie, P. (1985). The effects of instruction on main idea and question asking. *Reading-Canada-Lecture, 3,* 2.

Taylor, B.M., & Beach, R.W. (1984). The effects of text instruction on middle grade students' comprehension and production of expository text. *Reading Research Quarterly, 19,* 134-143.

Taylor, B.M., & Samuels, S.J. (1983). Children's use of text structure in the recall of expository material. *American Educational Research Journal, 20,* 517-528.

Van Dijk, T.A., & Kintsch, W. (1983). *Strategies of discourse comprehension.* New York: Academic.

Wong, B.Y. (1986). The efficacy of a self-questioning, summarization strategy for use by underachievers and learning disabled adolescents in social studies. *Learning Disabilities Focus, 2,* 20-35.

Wong, B.Y., & Jones, W. (1982). Increasing metacomprehension in learning disabled and normally achieving students through self-questioning. *Learning Disability Quarterly, 5,* 228-240.

CHAPTER **11** # *Framing: A Technique for Improving Learning from Science Texts*

Bonnie B. Armbruster

In this chapter Armbruster introduces framing—a procedure for helping students understand key ideas through identifying and using the structure of their textbooks. Frames help students visually understand main ideas and their interrelationships. Given that science texts rarely contain ready-made frames, Armbruster shows us how to construct frames based on sample elementary and secondary texts. She then describes how to use frames before, during, and after reading assignments.

Children generally have more difficulty understanding and remembering expository text, such as their science textbooks, than narrative text, such as stories in their basal readers (Berkowitz & Taylor, 1981; Spiro & Taylor, 1980). There are several reasons for this difference—inadequate prior knowledge of content, insufficient instruction in how to read expository text, too little practice reading expository selections, and a lack of interest or motivation.

According to recent research in learning from reading, children also have difficulty reading expository text because they are unaware of text structure. Therefore, they are unable to abstract the gist of the text or to relate the ideas to each other in a meaningful way. Research shows that students who are aware of text structure learn more from expository material than students who are not (McGee, 1982; Meyer, Brandt, & Bluth, 1980; Richgels et al., 1987; Taylor, 1980, 1985; Taylor & Samuels, 1983).

Research also shows that students can benefit from instruction in text structure. Effective instruction includes teaching students how to (1) use cues such as headings, subheadings, and paragraphs as indicators of text structure (Taylor, 1982; Taylor & Beach, 1984); (2) identify common expository text structures, including comparison-contrast, cause-effect, temporal sequence, problem-solution, description, and enumeration (Bartlett, 1978); (3) construct a diagrammatic representation of ideas and relationships in text (Armbruster & Anderson, 1980; Berkowitz, 1986; Dansereau et al., 1979); and (4) use provided graphic representations of ideas and relationships in text, called frames (Armbruster, Anderson, & Ostertag, 1987; Armbruster, Anderson, & Meyer, in press). The last study showed that over the course of one year, middle grade children learned more from their social studies textbooks if they worked with frames than if the teacher followed the suggestions in the teacher's manual. There is every reason to expect that frames would facilitate learning from science texts as well.

In sum, recent research suggests that awareness of text structure is important for learning from expository text and that instruction can help students identify and use text structures to their advantage. The purpose of this chapter is to introduce framing, a technique designed to increase comprehension and learning by helping readers identify and use the structure of their textbooks.

Frames: Definition and Examples

A frame is a visual representation of the organization of important content in a text. It depicts the main ideas of a text and the relationships connecting those ideas. Frames take a visual or graphic form for several reasons: (1) visuals are well suited to showing relationships among ideas; (2) many students can benefit from seeing information represented in a concrete, graphic form; and (3) a graphic can represent information succinctly, so it is a convenient way of summarizing a lot of information.

Although the term *frame* may be unfamiliar, the concept certainly is not. Familiar charts, tables, and diagrams may be frames. Many frames conveniently fit the form of a table or matrix, but frames need not take this form. Some semantic maps or graphic organizers also are frames. All of these graphics qualify as frames because they depict the main ideas or defining attributes of the topic or concept and show the relationships (e.g., descriptions, examples, comparisons, and contrasts) among

Figure 1
Sample Frame Formats

Biomes	Location	Climate	Plants	Animals
Tropical Rain Forest				
Grassland				
Desert				
Temperate Forest				
Taiga				
Tundra				

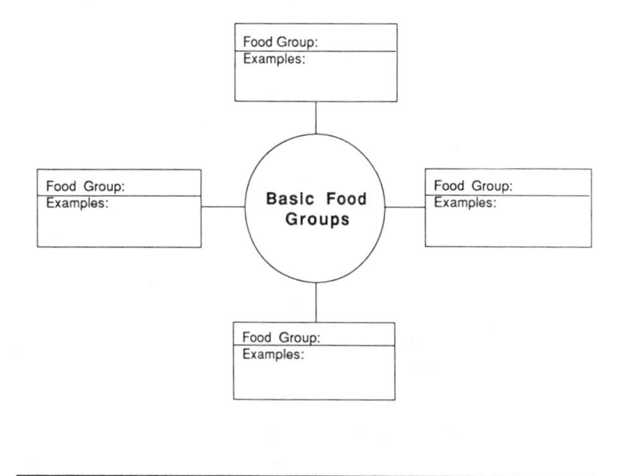

those main ideas. See Figure 1 for examples of frame formats.

How to Develop Frames

Science textbooks contain few ready-made frames, so teachers usually have to develop their own. Whether it is easy to develop a frame depends both on the developer's prior knowledge of the content and on how clearly and logically the textbook is organized.

When I develop a frame, I ask myself two questions: "What are the important categories of information associated with this topic?" and "How might these categories be subdivided?" The general procedure I use to generate frames is as follows. First, I use my prior knowledge of the content to develop a tentative frame. Next, I scan the chapter for clues to verify or revise my tentative frame. The cues I look for include headings, subheadings, introductions, summaries (or other indications of main ideas), and objectives. Then I read the chapter through to check my frame for accuracy and completeness. If my prior knowledge of the topic is particularly weak and the textbook is poorly organized, I sometimes have to consult additional sources—other textbooks, encyclopedias, or content matter experts.

Perhaps the easiest way to communicate how to develop a frame is to share examples of the process I used to create three science frames.

Example 1

The first example is a frame for a chapter entitled "Sources of Energy" in a fifth grade science textbook (Mallinson et al., 1985). Since this topic is familiar to me, I was able to quickly create a tentative frame. I expected the chapter to describe various kinds of energy sources, how they were obtained, what they were for, and the advantages and disadvantages of each. My tentative frame appears in Figure 2.

Next, I scanned the chapter to verify my tentative frame. The chapter headings were Energy from Fossil Fuels, Energy from the Sun, Energy from Water, Energy from Heat in the Earth, Energy from the Wind, and Energy from Living Things. The headings told me two things: (1) my terms *Geothermal* and *Solar* were not used in this textbook, and (2) I had left out the category of Living Things. Therefore, I needed to revise three columns of my frame. The revised frame appears in Figure 3.

Then I read the chapter to check for the accuracy and completeness of my frame. The three topics in my tentative frame indeed covered the important information discussed in

Figure 2
Tentative Frame for Sources of Energy

	Sources of Energy				
	Fossil fuels	Solar	Water	Wind	Geothermal
How is it obtained or produced?					
What is it used for?					
What are the advantages/ disadvantages?					

Figure 3
Final Frame for Sources of Energy

	Sources of Energy					
	Fossil fuels	Sun	Water	Heat in earth	Wind	Living things
How is it obtained or produced?						
What is it used for?						
What are the advantages/ disadvantages?						

the chapter, and there did not appear to be any additional topics. I concluded that my frame was accurate and complete.

Example 2

The second example is a frame for a chapter entitled "Machines at Work" from a third grade science textbook (Barufaldi et al., 1984). I don't know much about machines, but I figured I knew at least enough to make some hypotheses about how the authors of a third grade textbook might handle them. I thought a treatment of each machine would include a definition, a picture or diagram, a description of how it works, and some examples of how it's used. Figure 4 shows my tentative frame.

Next, I scanned the chapter. The headings were Lifting Things, Other Levers, Wheels with Teeth, Ramps, and Splitting Things. My confidence quickly waned as I realized that the chapter was not organized as I had surmised. Furthermore, I had difficulty telling from the headings just how it *was* organized. The headings were confusing because they were a mixture of objects (Other Levers, Wheels with Teeth, Ramps) and processes (Lifting Things, Splitting Things). I figured that Lifting Things probably included some mention of levers, and might include pulleys. I thought that Wheels with Teeth was about gears (a type of wheel and axle), and that Ramps was a type of inclined plane. Moreover, Ramps also might be the only acceptable vocabulary word at the third grade level. Splitting Things probably included a discussion of wedges. I could see that I would have to look further.

Then I scanned the Main Ideas and Objectives sections of the teacher's edition for each section, as well as the summary of main ideas ("What Did you Learn?") at the end of the chapter in the pupils' edition. From these sources I determined that only levers were discussed in Lifting Things. I also noted that definitions, diagrams, how they work, and how they are used were indeed categories of important information. In addition, the chapter included examples of complex machines that contain these simple machines. I thought this information might fit under the category of use, but that it would probably work best as a separate category (uses in other machines). For at least one machine (wedge), the chapter described examples of the "machine" in nature.

Given the uneven, inconsistent treatment

Figure 4
Tentative Frame for Machines

Machines

Definition:						
	Simple Machines					
	Lever	Wheel & axle	Pulley	Inclined plane	Wedge	Screw
Definition						
Diagram						
How it works						
How it's used						

Figure 5
Final Frame for Machines

Machines

Definition:				
	Simple Machines			
	Levers	Gears	Ramps	Wedges
Definition				
Diagram				
How they work				
How they are used				
Uses in other machines				
Examples in nature				

Figure 6
Tentative Frame for Waves

	Waves
Definition	
Characteristics	

of the different machines, I decided to read the chapter through to make sure I wasn't missing something. Finally, I was able to revise my frame. The final frame appears in Figure 5.

I felt rather uncomfortable with the frame because it didn't seem to be an accurate or complete representation of the content as I knew it to be, and because I realized that students would find if difficult or impossible to find information on some topics, but I felt that it was the best possible frame to use in teaching that particular textbook chapter.

Example 3

The third example is a frame for a chapter entitled "Waves" from a middle or junior high level textbook (Nolan & Tucker, 1984). With this subject matter, I was clearly out of my element. I figured the chapter would define waves and discuss their properties (I dimly remembered something about wavelengths and frequencies), but other than that, I was not able to hypothesize about the frame categories. My tentative frame is the very simple chart appearing in Figure 6.

Next, I looked at the chapter headings: The Nature of Waves, Changes in Waves, and The Doppler Effect. I guessed that The Nature of Waves might include both my categories of definition and characteristics, but I still could not surmise what Changes in Waves and The Doppler Effect would be about.

Then I scanned the chapter for objectives, which I found in both the teacher's and the pupils' editions. The objectives were quite informative. I was able to determine, for instance, that the chapter identified and defined two types of waves, discussed four characteristics of waves, and talked about how they are measured. It also listed three types of wave changes, explained how each worked, and gave the definition and uses of the Doppler Effect. Based on this information, I was able to develop the final frame that appears in Figure 7.

Finally, I read through the chapter to make sure I had included all of the important categories of information in the frame. I concluded that the frame for the chapter was accurate and complete.

Summary of Frames

As I hope these examples show, the process of generating a frame is similar to the process of producing an outline, except there is a greater emphasis on a meaningful visual representation. When developing a frame, a teacher must decide to what extent the frame will follow the textbook. Will it follow the textbook organization or impose an alternative organization on the content? I illustrated this dilemma in the discussion of Example 2, where I was torn between a frame based on my prior knowledge of the six simple machines and one based on the rather confusing and inconsistent organization of the textbook. The decision about which way to go will be a function of many variables, including the teacher's objectives, the students' abilities and interests, and time constraints.

Figure 7
Final Frame for Waves

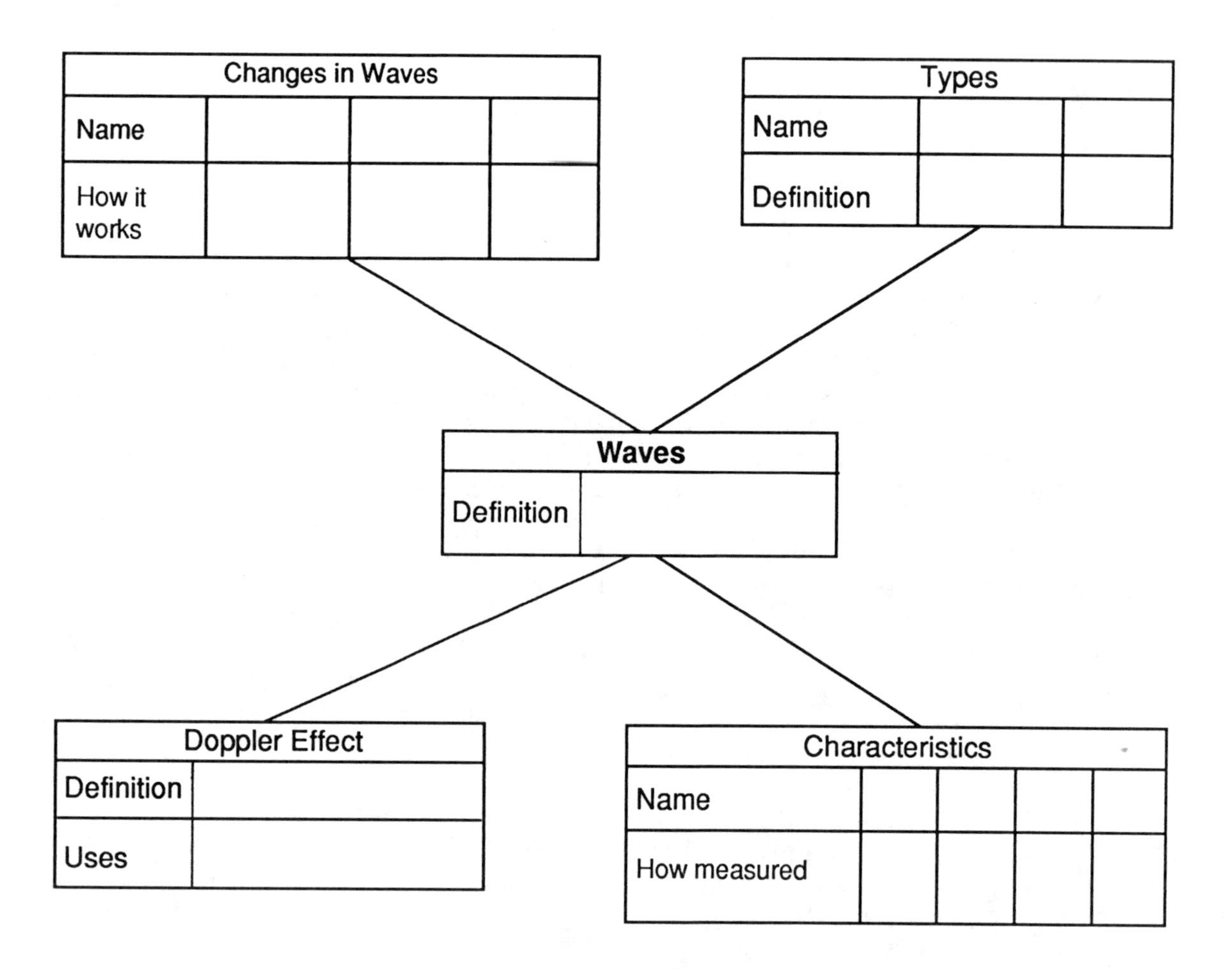

Teachers also need to decide the level of detail and explicitness of the frame. For example, in the "Waves" frame, I could have simply provided the headings—Definition, Types, Characteristics, Changes, Doppler Effect—without giving the additional information within each category. Or I could have indicated within the frame how the characteristics of waves are measured, which was also an important piece of information. Teachers' decisions about the level of detail to include in a frame will depend on what kind of information they want their students to know and how much help they feel students need.

Using Frames in Instruction

Framing forces teachers to think about the main ideas of the lesson and how they can be represented logically. Frames can be used to enhance science instruction before, during, and after reading assignments.

Before Reading

One way to use frames before a reading assignment is to have students develop the frame themselves. With the teacher's guidance, students could work through the steps previously outlined. Students would be encouraged to think about what they already know about the topic, form hypotheses about main ideas, and scan the text for information to confirm, refute, or expand their hypotheses.

Whether it is generated by a teacher or by a student, a blank frame (see examples in the figures) can serve as the basis for recommended prereading activities. These activities might include activating prior knowledge, building relevant background knowledge if such knowledge is absent, and setting a purpose for reading (Tierney & Cunningham, 1984).

First, a blank frame can be used to help students think about what they already know about the topic while providing focus categories for student brainstorming. The frames then can be used to record prior knowledge, hypotheses, and predictions about upcoming content.

Second, a blank frame can be used to organize instruction about important background information. Teachers may compare what is known about the target content of the frame with information provided in the text. If students' lack of prerequisite knowledge would inhibit their understanding of the textbook, teachers can use the frame to organize prereading instruction.

Third, a blank frame sets a purpose for reading. The frame shows students what they should learn and remember from what they read; it acts as a special study guide, with its own implicit learning objectives.

During Reading

Both students and teachers can use frames as the students read their science textbooks. Students can use frames as a structured format for taking notes—a difficult skill to acquire and teach. Younger and poorer readers usually do not have well-developed strategies for taking notes (Armbruster, Echols, & Brown, 1982). One reason these readers are not strategic note-takers is that they do not know which information is important to record. A blank frame shows students which information is worth recording and, by providing limited space, forces students to paraphrase ideas concisely. Recording information in a blank frame also keeps students actively involved in learning from reading. Students must actively process the text to decide which information belongs in the frame and where.

Teachers can use frames during reading to teach and develop higher order thinking skills. For example, students practice classifying information as they sort ideas into cells of the frame. They can develop the skills of making inferences and drawing conclusions by trying to figure out information required by the frame that is not explicitly (or even implicitly) given in the textbook. The skills of comparing and contrasting can be fostered by having students compare and contrast information across the rows or columns of matrix frames.

After Reading

Again, both students and teachers can use frames after the reading assignment. Students can use a completed frame as a summary of the text—a useful study reference in preparing for a test. They can use a blank frame as a means of evaluating their learning. That is, they can ask themselves questions based on the frame categories, or they can see how well they can fill in the frame from memory.

Teachers can use frames to help students review information or to evaluate learning. Frames can be used in evaluation in two ways. First, teachers can use the frame categories as the basis for standard questions (e.g., short answer, multiple-choice, matching). For the frame in Figure 2, a question might be "What are the advantages and disadvantages of using energy from the sun?" Second, teachers can evaluate learning by having students fill in a cloze frame (partially completed frame with some cells left for students to complete).

Using Frames in Textbook Evaluation

Textbook adoption committees are charged with the task of evaluating and selecting new textbooks. Unfortunately, the people serving on these committees probably have never been trained to evaluate or select materials for classroom use (EPIE, 1976). Therefore, they tend to rely on publishers' sales presentations and locally produced checklists. These checklists often contain an item that refers vaguely to the quality of the textbook's organization. Framing can help adoption committees define and standardize evaluation criteria for text organization.

In our work with frames, we have found a direct relationship between the ease of framing and organization of the textbook. If the text is poorly organized, we have difficulty generating frames. Textbook adoption committees can use this relationship in making their evaluations.

A practical suggestion for these committees is to choose a target topic that is considered important in the school's curriculum and that is covered in all candidate textbooks. The content probably should be no longer than a single chapter. Using the suggested procedure, the committee should try to construct a frame for the target content of each book. The textbook that yields the most accurate and complete frame with the least effort is probably the best organized textbook, at least for this (presumably representative) topic.

References

Armbruster, B.B., & Anderson, T.H. (1980). *The effect of mapping on the free recall of expository text* (Tech. Rep. No. 160). Urbana, IL: University of Illinois, Center for the Study of Reading.

Armbruster, B.B., Anderson, T.H., & Ostertag, J. (1987). Does structure/summarization instruction facilitate learning from expository text? *Reading Research Quarterly, 22,* 331-346.

Armbruster, B.B., Anderson, T.H., & Meyer, J.L. (in press). Improving content area reading using instructional graphics. *Reading Research Quarterly.*

Armbruster, B.B., Echols, C.H., & Brown, A.L. (1982). The role of metacognition in reading to learn: A developmental perspective. *Volta Review, 84,* 46-56.

Bartlett, B.J., (1978). *Top level structure in an organizational strategy for recall of classroom text.* Unpublished doctoral dissertation, Arizona State University, Tempe, AZ.

Barufaldi, J.P., Ladd, G.T., Moses, A.J., Schneider, H., & Schneider, N. (1984). *Health science, level 3.* Lexington, MA: D.C. Heath.

Berkowitz, S.J. (1986). Effects of instruction in text organization on sixth grade students' memory for expository reading. *Reading Research Quarterly, 21,* 161-178.

Berkowitz, S.J., & Taylor, B.M. (1981). The effects of text type and familiarity of the nature of information recalled by readers. In M. Kamil (Ed.), *Directions in reading: Research and instruction.* Washington, DC: National Reading Conference.

Dansereau, D.F., Collins, K.W., McDonald, B.A., Holley, C.D., Garland, J., Diekhoff, G., & Evans, S.H. (1979). Development and evaluation of a learning strategy training program. *Journal of Educational Psychology, 71,* 64-73.

EPIE Institute. (1976). *Research findings: NSAIM, two years later.* New York: EPIEGRAM.

Mallinson, G.G., Mallinson, J.B., Smallwood, W.L., & Valentino, C. (1985). *Silver Burdett science (grade 5).* Morristown, NJ: Silver Burdett & Ginn.

McGee, L.M. (1982). Awareness of text structure: Effects on children's recall of expository text. *Reading Research Quarterly, 17,* 581-590.

Meyer, B.J.F., Brandt, D.M., & Bluth, G.J. (1980). Use of top-level structure in text: Key for reading comprehension of ninth grade students. *Reading Research Quarterly, 16,* 72-103.

Nolan, L.M., & Tucker, W. (1984). *Health physical science.* Lexington, MA: D.C. Heath.

Richgels, D.J., McGee, L.M., Lomax, R.G., & Sheard, C. (1987). Awareness of four text structures: Effects on recall of expository text. *Reading Research Quarterly, 22,* 177-196.

Spiro, R.J., & Taylor, B.M. (1980). *On investigating children's transition from narrative to expository discourse: The multidimensional nature of psychological text classification* (Tech. Rep. No. 195). Urbana, IL: University of Illinois, Center for the Study of Reading.

Taylor, B.M. (1980). Children's memory for expository text after reading. *Reading Research Quarterly, 15,* 399-411.

Taylor, B.M. (1982). Text structure and children's comprehension and memory of expository material. *Journal of Educational Psychology, 74,* 323-340.

Taylor, B.M. (1985). Toward an understanding of factors contributing to children's difficulty summarizing textbook material. In J.A. Niles (Ed.), *Issues in literacy: A research perspective* (pp. 125-131). Rochester, NY: National Reading Conference.

Taylor, B.M., & Beach, R.W. (1984). The effect of text structure instruction on middle-grade students' comprehension and production of expository text. *Reading Research Quarterly, 19,* 134-146.

Taylor, B.M., & Samuels, S.J. (1983). Children's use of text structure in recall of expository material. *American Educational Research Journal, 20,* 517-528.

Tierney, R.J., & Cunningham, J.W. (1984). Research on teaching reading comprehension. In P.D. Pearson (Ed.), *Handbook of reading research* (pp. 609-655). White Plains, NY: Longman.

CHAPTER 12 *Tools for Learning Science*

Shirley Harrison

Harrison takes us into her classroom to explain how she uses two-column frames with her ninth grade earth science students. She walks us through four procedures that help her students think critically about science: problem-solution, theory-evidence, liked-disliked, and question-answer strategies. In each case she explains how to use these tools for analyzing information, defending arguments, and writing about scientific topics. Elementary teachers will find her ideas readily adaptable for younger students.

Science is boring. Our text is boring. Lectures are boring. The mere mention of this six-letter word raises the hair on the back of my neck and makes me smart as if I had been stung with the label of—heaven forbid—a boring teacher! Boring science texts, lectures, and teachers are not unique. Many of us as students uttered these same words.

Teaching my junior high students how to learn science is in itself a science that requires careful consideration and use of the scientific process. Students need a variety of activities to trigger their curiosity. To maintain their interest they need techniques or tools that will allow them to continue learning beyond the science classroom. Using tools enables them to learn better and consequently feel less frustrated and bored in my classroom.

Much like the Galapagos woodpecker finch, which uses a spine or twig to pluck insects from a cactus, we use learning tools in our daily lives. In the science classroom, knowing how to think, read, and write are necessary tools for success. But frustration and apathy result when we tell our students to learn without first showing them how to do so. These observations have led me to develop tools for helping my students learn.

I call these tools two-column frames. These organizational schemes are similar to those described by Armbruster (this volume), but they are not based on the specific structure of the text. They are more generic. In fact, the same frame format can help students organize science information from newspapers, videos, library books, magazines, filmstrips, and texts.

I use four variations of these tools: the problem-solution, theory-evidence, liked-disliked, and question-answer frames. For all these variations, modeling is essential. I introduce each by demonstrating to students how I construct the frame. Students practice using frames in class before doing them on their own.

Problem-Solution Frame

When introducing the problem-solution frame, I begin with something familiar. "Who has a problem that is really irritating?" After a few moments of silence, followed by surprised looks and quizzical exchanges, numerous brave hands seek attention. My first step is to write the word Problem on the board and choose a student who seems desperate to discuss his or her problem. I ask the student to state the problem, which I then proceed to write on the board.

"I always get bad grades in school," the student says. Other students join the discussion. After writing Result I ask, "What happens because of this problem?" The student replies, "My parents are mad at me." Since parent frustration is common during this adolescent age, several others chime in. "My parents ground me. I can't use the car. I can't go to the game." These answers go on the board.

Now that all are involved, I change the momentum by asking the entire class: "What do you think causes these problems?" I write Cause beneath Result and again fill the chalkboard with students' responses. "The work is too hard." "I don't like the teachers." "I don't do homework."

In the third step, I ask, "What might be a Solution to the original problem of bad grades?" The students' numerous solutions fill the rest of the chalkboard (see Figure 1).

As we reflect on the responses to the key words—Problem, Result, Cause, and Solution—I ask how this same process might work in earth science. After a brief discussion, I give the students a photocopy of a recent article related to the science topic we are studying. I make a transparency of the article, and after I introduce key vocabulary, students read the selection silently. Then together we reread and discuss information related to the four key areas. I show them how to underline selectively and make marginal notes about which of the key areas the underlined information relates to. The students use their underlines and marginal notes for completing the frame.

For one lesson, I assigned a newspaper article describing Americans as having a poor background in science. After reading and underlining key information, the students took notes on the problem, result, cause, and solution. In this article, the author did not directly state a solution, so students came up with their own solutions, as recorded in Figure 1.

In addition to organizing the content around the four areas, we also talked about several important vocabulary words and listed brief definitions of these words in our notes. Notice that students use key words and phrases rather than complete sentences in their notes. This not only saves time but also discourages plagiarism, particularly when using notes to write reports.

The following day, students talk about the information in their notes. Because I encourage divergent thinking and variation in responses, I provide opportunities for students to explain their responses. Next, I demonstrate how to use the information in their notes to write a paragraph. (Problem-solution notes evolve naturally into writing.) We begin the paragraph with a problem statement followed by information about the result, cause, and solution. A sample student paragraph appears below.

Student Paragraph

> Americans are a technologically confused society. The result is they are not informed on technological issues. Forty percent or more of Americans believe flying saucers are real, that rockets change the weather, and that numbers bring good luck. The cause of this seems to be students are graduating unprepared to grasp even day-to-day issues. Some solutions to the above problem might be to require students to take more science, emphasize technology in school, and educate teachers in science.

After doing the entire process together, I nonchalantly hand out three or four different magazine or newspaper articles on the same

Figure 1
Problem-Solution Frames

The teacher introduces the frame through a class discussion about a common student problem.

Problem:	I always get bad grades.
Result:	My parents ground me. I lose my freedom—car, entertainment. I lose my self-esteem.
Cause:	I don't like teachers. The work is too hard. I don't do homework.
Solution:	Pay attention during class. Take a study skills class. Do homework on a consistent basis. Change lab partners.

The teacher then uses the same format with a science selection. In the following example, students read, underlined, and took notes on a magazine article about Americans' poor background in scientific technology.

Problem:	Most Americans do not understand new technology.
Result:	They're not informed on technological issues. Forty percent or more of Americans believe flying saucers are real, rockets change weather, and numbers bring good luck.
Cause:	Graduating students are unprepared to grasp day-to-day issues.
Solution:	Require students to take more science. Emphasize technology in school. Educate teachers in science.

controversial earth science topic. The articles describe the same issue but from different points of view. Each student reads and takes notes using the problem-solution format. A lively debate occurs when the class discovers one topic has several problems, results, causes, and solutions. Students quickly learn to read critically. Problems never have simple answers—especially in science.

Theory-Evidence Frame

The theory-evidence frame helps students begin to think like scientists. I spend the first part of the school year on the scientific process and how theories, laws, and facts develop. Integral to teaching the scientific process is analyzing written material according to theoretical assumptions and evidence used to support a

Figure 2
Theory-Evidence Frame

Theory	Evidence Author Uses to Prove It
1. What is the theory? (Students summarize the author's theory.)	1. First evidence. (Students write or describe here.) 2. Second evidence. (Students list the evidence, usually in the same order as presented in the article.) 3. Third evidence. (Students fill in any remaining evidence.)
2. What do you think? (Here students state opinions and hypotheses of their own.)	

Figure 3
Theory-Evidence Notes

Theory	Evidence to Prove It
1. What is the theory? Large asteroid about six miles in diameter crashed into earth, killing the dinosaurs.	1. Impact caused rise in earth's temperature; dust in atmosphere. 2. Dust blocked off light; plants died, no food for dinosaurs. 3. Extinction was quick.
2. What do you think? Seems pretty far-fetched to me. No evidence of spot on earth where asteroid hit.	

theory. It is important for students to realize that scientific knowledge constantly changes. Often new evidence and conclusions can be conflicting and difficult to understand. In most situations, authors present their thesis or theory in the opening paragraphs, with the rest of the paper dedicated to presenting evidence supporting the theory.

I begin by modeling the strategy with a short article. The process works best when students have their own photocopies. With the use of an overhead projector, we talk about how authors organize their writing. Then we underline the theory and supporting evidence. I demonstrate how to look for these components and how to underline and write brief notes in the margins.

I then give students a handout similar to

the example shown in Figure 2, and together we develop notes. Notice that in addition to notes on the theory and evidence, there is space for generating a conclusion.

After developing one or two theory-evidence frames together, students begin reading and analyzing articles on their own. I allow class time for discussion to give students opportunities to explain their answers. They often find that the author's theories lack enough convincing evidence. I always allow students to derive their own conclusions as long as they have adequate evidence. Students can comprehend difficult magazine or newspaper articles when they learn this simple theory-evidence model.

In the example in Figure 3, a student took notes using the theory-evidence format while reading an article entitled "Death Star" by David Quanman from the September 1984 issue of *Outside* magazine. The frame helped the student organize and evaluate the information presented.

As with the problem-solution frame, this activity naturally leads into writing summary paragraphs. Notes describing the theory are woven into an opening sentence, and notes from the evidence column become supporting sentences. To help with the transition from notes to writing, I sometimes provide a paragraph frame to help students organize their first draft (Figure 4).

Once students know how to organize a theory-evidence paragraph, they rely less on my paragraph frame. I encourage them to vary from the paragraph frame as long as they describe the theory along with supportive details. Not only are their paragraphs easy to correct, but I know immediately if they understand important scientific theories. A sample piece of student writing on the "Death Star" article appears below:

Student Writing Sample

The author concludes that dinosaurs disappeared because of a giant asteroid that hit the earth. First, the author says that the asteroid probably caused a rise in the temperature of the earth. The rise in temperature may have caused the dinosaurs to die. Second, when the asteroid hit, it created a lot of dust and particles. The dust blocked out the sun so plants could not grow. The dinosaurs probably starved to death. Third, the asteroid caused dinosaurs to die quickly.

I think the author's theory might be true. Dinosaurs would starve to death if they

Figure 4
Student Paragraph Frame

The author's theory in this article is ____

____. The first evidence the author

presents is ____. Second, ______________

__________. Third, ____. I think that ____

________________________.

didn't have any food. I am not in total agreement with the theory because it seems like there should be a great big hole someplace on the earth where the asteroid hit the earth.

The same theory-evidence tool works well when we are studying a controversial science concept. I often divide the class into three or four teams, each doing a theory-evidence study sheet and paragraph on the same concept but with different authors and theories. When students finish writing their summaries, they present them to the class and answer questions from other teams. These presentations and questions create lively classroom discussions. Moreover, students learn that several different theories can be equally convincing.

For example, we recently debated about the disappearance of dinosaurs. I used three different articles, each supporting a different theory. Teams then used their notes for an oral debate. Following the discussion, they wrote persuasive paragraphs defending the most viable theory. The theory-evidence frame works well for junior high earth science students. They become critical readers and writers, and learn to think through scientific issues.

Liked-Disliked Frame

Junior high students have opinions on everything. The only problem is that they seldom provide a rationale for what they believe. Therefore, I decided to use in a productive way what students do naturally. My scheme, which I call the liked-disliked frame, fits perfectly with what we know about learning science (see Finley, Armbruster, Baker, and Santa & Havens, this volume). In order to learn, students need to combine their prior knowledge with incoming information. We can't fill our students with knowledge; we can lay information before them, but they must act on it.

A good way to bring students' background knowledge and personal reactions into the reading or listening situation is to ask them to form opinions about science information. This is what the liked-disliked frame does. This tool encourages students to become more actively involved in forming opinions about ideas from films, lectures, and reading assignments. They then use their notes for discussion and writing. In the example in Figure 5, my students watched and responded to "The Bridge," an award-winning film on state parks produced by the Montana State Fish, Wildlife, and Parks Association.

Figure 5
Liked-Disliked Frame

Earth Science Film Evaluation

The title of the film is ____The Bridge.____.

What I Liked About the Film:	What I Disliked:
The way the narrator used metaphors and similes.	Used words I didn't understand.
The buffalo jump and the hunters. Place to escape.	The first part about the bridge didn't apply to the rest of the film.
Showed contrasting environments.	

This simple two-column tool focuses student attention on the task. Students take activities such as viewing films more seriously when asked to take brief notes and write about their reactions. As they become more confident they begin to create their own focus for taking notes and for writing paragraphs from those notes. They take charge of their own learning. Boring—that nasty word—reaches my ears less and less frequently.

Question-Answer Frame

Our earth science text is packed with concepts, and some parts are not written clearly. Many students become frustrated and won't read assignments without help. Even my good students have difficulty organizing the numerous and complex ideas presented in the text.

The question-answer frame is a partial solution to this problem. Students organize text information into two columns. On the left, they write questions reflecting the key ideas presented in the selection. On the right, they write answers to these questions. When they complete the frame, they fold their paper in half lengthwise and ask themselves the questions without looking at the answers on the other side. Then they turn the paper over to verify their answers.

Even though this format is simple, I must model the strategy before students can successfully create their own question-answer frames. As with the other two-column activities, I use an overhead projector and give students photocopies of their own. Taking one section or topic at a time we work through the chapter. First, we examine the author's clues to main ideas such as headings, bold or italic print, rhetorical questions within the text, chapter objectives, and end-of-chapter questions. Then I have students read about a topic, looking for main ideas as they read. When they're done, we discuss the main ideas and create questions. We follow the same procedure with the next section. Finally, students reread the chapter and answer the questions on their own. The following day, we use class time to discuss their answers.

After modeling this procedure, I no longer make photocopies of the chapters, but if necessary, I continue to demonstrate. When introducing a new chapter to the class, we talk about the content so that students know what to include in their notes. If the reading is particularly difficult, I make a transparency and we walk through the chapter together, writing our questions as part of our discussion. Students then reread the chapter and answer the questions on their own. As the year goes on, they take more and more responsibility for creating their own question-answer notes.

After the initial learning of the question-answer frame, I vary the assignment to perk students' motivation and to prevent them from using just vocabulary/definition type questions or the boldfaced words. Students love being challenged rather than doing the same assignment each time. I might, for instance, ask them to write a specific number of questions and an answer key. They exchange questions with classmates the next day. Then they return their written answers to the question authors for correcting. These student-developed quizzes motivate thorough reading and help students become aware of poorly written questions. They complain that some of their peers' questions are unclear.

This awareness leads naturally into lessons on how to write questions. Using a transparency displaying a page from their science text, I write a knowledge-level question from a sentence in the text, and we talk about how we need to know only simple facts to answer this type of question. I point out that over 90 percent of the questions they wrote are knowledge-level questions. (These take less reading and are easiest to write. My clever students know how to complete assignments by doing the minimum amount of work.)

After reading my example question, I ask my students to make it more difficult. After much discussion, they usually agree that a good question needs to be drawn from more than one sentence in the text. I ask if they can make up a question that uses information from three or four sentences. I call these comprehension questions. After developing several comprehension questions together, I ask the students to write five knowledge and five comprehension questions on their next reading assignment.

When they bring their questions and answer keys the next day, they are more aware of questions that demand a fuller understanding. After several reading assignments, I ask for more difficult questions requiring application, analysis, and synthesis of information. I motivate students by awarding more points to more difficult questions and by offering extra points when they label the types of questions I use on tests.

Integrating instruction on questioning with the question-answer frame builds students' thinking/questioning skills. By the end of the school year, they become quite smug about their ability to recognize and write difficult questions. They critique my tests and complain if I give too many analysis and synthesis questions. I am delighted when they know enough to be critical.

Conclusion

The two-column frames stimulate reading, thinking, and writing. They give students focus and direction for organizing information from science.

Even more important, the word *boring* occurs less frequently when students know how to tackle scientific information. It is easy to be bored when you feel insecure. Earth science naturally invites curiosity and has even stronger appeal when students have ways to learn and react personally to information.

These four frames provide variety, which prevents apathy and increases students' tool chest of learning. My students can use these tools in and outside the classroom, now and for the rest of their lives. Student success depends on teacher enthusiasm and modeling to ensure that they feel comfortable as learners. But the hallmark for me as a teacher is when my students begin showing me tools for learning that they have devised. Perhaps I've helped launch their careers as independent learners.

CHAPTER 13 *Learning Through Writing*

Carol Minnick Santa
Lynn T. Havens

Santa and Havens describe how science teachers in their district use writing to help students learn science. They begin by explaining why writing is essential to learning both the content and the processes of science. Then they present secondary classrooms where teachers demonstrate how they have integrated learning logs, scientific reports, and written explanations of scientific phenomena into their teaching. Teachers will find the ideas for writing in science adaptable to any grade level.

The science teachers in our district have a special view of the world. Their knowledge helps them appreciate and understand nature. When they hike the jagged peaks that surround our mountain valley, they know each tree and wildflower. They understand how glaciers carve valleys, and can tell true stories about secret homes of wildlife. Their background empowers them to see nature's extraordinary pen.

As curriculum facilitators in our district, we are fortunate to collaborate with science teachers. For the past 6 years, we have worked together exploring ways to help our students learn science. We want them to know both the content and process of science so they can more clearly view, understand, and enjoy their world.

Science teachers feel responsible for covering content. Science in particular builds on a rich base of prior knowledge that lays the foundation for later courses. Yet there is far more to science than content. Our students also must learn to observe and think about scientific phenomena. Science processes enable them to learn and think through issues long after they leave the supportive confines of the classroom.

One way to meet both content and process objectives is through writing. When first thinking about incorporating writing into our science classes, we asked ourselves, Won't writing take precious time from protozoa, protein synthesis, and phenotypes? Isn't writing the English teachers' responsibility? How can scientists teach writing? Science teachers aren't familiar with dangling participles, split infinitives, and five paragraph papers.

To address these issues, we combined our personal ideas about writing with our theoretical knowledge about reading and learning. We know that writing works as a powerful teaching tool. It reaches the heart of learning by penetrating the external shell of memorized fact and superficial understanding. As teachers, we write to understand. If we can explain things to ourselves and to others, we can claim new knowledge as our own. From a personal perspective, writing makes sense.

In addition, writing provides an avenue for us to practice what we know about background knowledge, active learning, organization, and metacognition. Science teachers understand the power of background knowledge. Our students have difficulty understanding labs or complex ideas in their reading unless we first teach necessary concepts. We do this by finding out what students know about a topic and then teaching necessary concepts before students read or conduct laboratory activities. Integrating new information with background knowledge is fundamental to science learning, and writing can help with this process. Writing about a topic before reading enlists prior knowledge, making it ripe to incorporate with new information. Writing afterwards bridges the new information with the old.

Writing encourages active involvement in learning. Effective learning is not something we can do for our students; it requires initiative. Too often students remain passive, waiting for teachers to fill them with knowledge. It is impossible to remain passive as a writer, however. When writing about observations, or about information in a reading assignment, we necessarily participate in the learning process.

Writing forces organization. It helps us see clusters of information and hierarchies of ideas. As we build systems of organization, we make new information our own.

Writing helps students become metacognitive. Baker (this volume) describes the concept of metacognition, explaining how good readers monitor their comprehension. Good readers know when they understand, and what to do when they do not. Writing helps students gain this awareness, in part by providing a means of self-measuring knowledge. We cannot write about something if we do not understand it.

Because writing makes sense personally and theoretically, we decided to implement writing in science. We started cautiously, adding a few ideas each year and evaluating their effectiveness informally. We began to observe students' success on exams based on content they had written about in class. We no longer question the validity of writing as a tool for learning science.

We use a variety of writing assignments in our classes. Three of the most successful types of assignments are learning logs, scientific reports, and written explanations of scientific phenomena.

Learning Logs

What began as a cautious experiment is now a critical component of many science classes. Students entering biology or earth science express surprise when asked to keep a learning log. Most think such writing belongs in the English class. Before long, however, they understand why expressive writing helps them learn science, particularly the difficult concepts.

During the first week of school, our science teachers usually ask their students to purchase a loose-leaf notebook with dividers for separating lab work, class notes, and the learning log. Teachers keep their own logs, which they use for modeling.

Teachers often introduce logs by explaining why writing is essential for learning. They discuss their own experiences in keeping a log and show students their entries. They explain that at times, their entries may be questions about vocabulary, difficult concepts, or lab results. In other instances, their entries may be phrases explaining the text or lab observations. Entries often are messy, with thoughts scratched out and arrows indicating added information. Teachers show how their rough recordings of observations and questions lead to a conclusion. Expressive writing is their thinking written down. One teacher told the class,

"Because I write, I know where I stand. Writing provides a status of my thoughts, and forces me to grapple with what I know and what I don't know." Such awareness is the essence of metacognition and provides the key to continual learning.

After discussing their logs, teachers often introduce students to expressive writing as part of a reading assignment. For example, teachers ask students to examine the title, subheadings, and chapter questions before reading and then brainstorm in their logs what they already know (or think they know) about the topic. Pairs of students read their entries to one another. After reading the text selection, they write another entry focusing on what they learned and noting misconceptions in their initial entries. Then they read their new entries to their partners.

Recently students in a biology class read about flower reproduction. The following example shows one student's pre- and postreading entries.

Before-Reading Entry

> In this chapter I am going to learn about flower reproduction. I know that flowers have male and female parts. I think that these parts are in the inside of the flower. To see them you have to pull aside the petals. I think petals probably protect the reproductive parts, but I am not sure. I remember something about separate flowers for male and females, but I think many flowers have both parts on the same flower. I'm pretty sure you need to have at least two plants before they can reproduce.

After-Reading Entry

> I learned that stamens are the male parts of the flower. The stamen produces the pollen. The female part is the pistil. At the bottom of the pistil is the ovary. Plants have eggs just like humans. The eggs are kept in the ovary. I still am not sure how pollen gets to the female part. Do bees do all this work, or are there other ways to pollinate? I was right, sometimes male and female parts are on separate flowers. These are called incomplete flowers. Complete flowers have both male and female parts on the same flower. I also learned that with complete flowers just one plant can reproduce itself. So, I was partially wrong thinking it always took two plants to reproduce.

The teacher concluded the lesson by asking students to write about the learning log strategy. (We call these reflective entries process discussions.) The teacher asked, "What did writing about your ideas before and after reading do for you as a learner?" The student's response follows.

Process Discussion Entry

> I felt more interested in reading about flowers because I thought about it before I read. I was surprised I knew as much about flowers before reading. Writing helped me put down what I already knew. It made me more curious about reading, because I wanted to know if I was right. Knowing that I was going to have to write when I finished reading made me read more carefully. I got more out of my reading by writing. I wasn't very happy about doing it at first, but it helped.

Next the teacher had volunteers read their process discussion entries, which sparked a lively discussion about learning. It is essential that students understand why informal writing is a powerful tool for learning and why an assignment like the one just described is so effective. If students know why writing works, they are more likely to use it as one of their own learning strategies.

With this goal in mind, class discussion focused on how writing helps link background knowledge to the topic, generates active learning, and leads to self-monitoring. Such discussions enable students to understand themselves as learners. Most students have only vague ideas, if any, about the usefulness of writing. By providing opportunities for writing and talking about why it works, teachers often can convince students to use writing on their own.

In addition to using learning logs for process conferences and as part of pre- and post-reading activities, teachers use them extensively in conjunction with scientific observations. Learning to observe is one of the most basic and important acts of science. Students must use all of their senses as they notice how things may be alike or different and how things change over time. Writing provides the feedback and direction needed to learn the skills of observation. We want our students to find nuances in nature, to hypothesize and explain their observations. Writing about what they see forces them to organize their observations and allows them to see even more. Learning logs are vehicles for seeing, and for developing and recording ideas. As students write their observations, they discover meaning in what they see.

Learning logs also provide a forum for asking questions. In science, problem solving and question asking are stressed. Yet, who asks most of the questions in our classes? Often we sit behind our desks and begin the inquisition. A sprinkling of students become actively engaged in answering our questions, but does this system create independent learners? During the brief hours we have our students, we must show them how to become self-learners and self-questioners. Learning logs can help bridge the gap from passive to active learning and thus foster independence.

Learning logs inspire precise observations, which naturally lead to questions. For example, in the following situation students were observing a flower and recording their observations.

Observation Entry

> Right below the outside of the flower the stem thickens into a little case or holder for supporting the flower. It looks almost like a crown. From this crownlike form is a ring of petals. The petals are in two layers. The lower petals are greener and look more stemlike than the upper petals. I wonder what use all of these petals have? I know they might attract bees for pollination. It could be that their function is more for protection of the fragile internal parts of the flowers.
>
> Next, I see a ring of tiny petal-like parts in the middle of the flower. There are four of these, and each tiny petal has two sharp spires on each side. I wonder if these are the stamens and pistils? They look very different from the surrounding petals—far more delicate.

Notice how easily observations led to questions, especially when students committed their observations to paper. This student's log set the stage for motivated learning. He wanted to know what his observations meant. He generated his own hypotheses, or purposes for reading.

While we favor using logs as observation tools, we also use them in other ways. For example, after students watch a film, complete an experiment, or listen to an oral presentation, we allow time for writing. We ask, What struck you as important in this session? What do you want to remember? Teachers often participate in these assignments, writing in their logs along with the students. Then they can read their entries aloud and invite students to read theirs.

Logs provide opportunities to write informally and to explore content. The idea is to write without fear of teacher interrogation or a hemorrhaging red pen. The writer is always the primary audience. We never grade logs, although we usually give points for completion. Junior and senior high school students need "point" motivation.

We do recommend reading the logs. Teachers usually find time to read several during the class period. Reading the logs gives teachers a better understanding of their students' knowledge of science. Students' entries provide important data about the teacher's presentation; their questions and confusions guide the teacher's instructional planning.

One teacher asks his students to leave space for him to write. Student entries trigger his own ideas, and he can't resist writing back. Sometimes his responses are questions, sometimes clarifying comments about their observations, but more often he nudges with ideas for further exploration. This personal dialogue provides the individual attention so often missing with high school students.

Student reactions to learning logs provide the real test of their effectiveness. Many students begin with understated interest, but then opinions begin to shift and initial reservations change to enthusiasm. (We have yet to find a log in the garbage can.)

Having students compare their entries from the beginning of the year with those at the end of the year can provide further motivation. Most of their early entries are brief, tersely written statements directed more to teachers than to themselves. They are tightly controlled and impersonal; there is no recklessness or testing of ideas. At the end of the year, entries are longer and messier. Logs become more of a transcription of thinking. Thoughts are crossed out; new ones take their place. Questions become more frequent as students explore their own observations and thinking. Logs evolve into tools for exploration.

How can teachers inspire this kind of change? First, students must feel safe for logs to be effective. Entries must be spared from red pen intimidation and criticism. Second, students must write frequently. Provide time in class for writing. Keep your own log and model your own writing as students write. If students see teachers taking writing seriously, they begin to internalize their own need to write. Third, write back to students in their logs. When we converse with students through writing, their responses become more exploratory. Our encouraging comments make logs safe.

Learning logs can become ongoing tools in the science classroom. In many cases, observations and comments contained in learning logs become important ingredients in the more formal scientific report.

Scientific Reports

We want our students to think like scientists. They need to know how to define a problem, how to analyze the problem, and how to formulate conclusions. Conducting and writing about experiments is the best way to acquire this knowledge. Recall for a moment some of your college science classes. You may forget the details of advanced biology and biochemistry, but you probably have not forgotten how to investigate scientific questions. We know how to think through issues because we have experienced the scientific method with hands-on laboratory experiments.

Scientific thinking is akin to solving mysteries. We tell our students that being a scientific detective involves a special way of thinking and experimenting. It includes observing, questioning, hypothesizing, experimenting, analyzing, and concluding. We can't teach these process skills without involving students in writing. Scientific problem solving and scientific report writing go hand in hand. If students write about the steps in the process, they can take hold of their thinking. Writing glues thinking on paper.

We begin teaching the scientific method with simple problems the class works through together. Along with this direct instruction, students use written guidelines for conducting an experiment (Figure 1). Responses to the guidelines become ingredients for a laboratory report.

The teacher begins with what the class already knows about a topic. The class generates one or two problems or questions to investigate. Students recall knowledge related to the topic, and together they write educated guesses or hypotheses as explanations. Then they brainstorm ways to test the hypotheses,

Figure 1
Laboratory Guidelines

1. *Purpose.* Why are you doing this lab?
2. *Problem.* What problem are you investigating?
3. *Hypothesis.* What do you think the outcome of this lab will be? Think about what you already know about the topic and make an educated guess about the outcome.
4. *Materials.*
5. *Procedure.*
6. *Data or Results.* Draw, record, and chart all detailed observations noted from the procedure above.
7. *Analysis or Conclusion.* Reread your problem statement. Did your results resolve the problem? Explain why, using data from the experiment. If the results did not resolve the problem, explain why.
8. *Class Conclusion.* Final conclusion after class discussion. Modify your earlier conclusion if necessary.

develop an experiment, and list procedures. Once they have completed the experiment and recorded the results, they go back and write answers to their original questions. Did the investigation answer the questions? If not, they come up with alternative ways to examine the issues. Finally, together they draft a laboratory report. Teachers work through many such whole-class experiments until students begin to feel comfortable with the procedure.

Figure 2 provides an example of a student's laboratory notes, which follow the laboratory guidelines. In this experiment, ninth graders examined factors influencing the density of objects. Students investigated two hypotheses regarding the effects of size, shape, and composition.

The students write scientific reports from their laboratory notes. These reports have four sections: purpose, materials and procedure, results, and conclusions. The purpose section includes information from Steps 1, 2, and 3 of the laboratory guidelines; the materials and procedure section includes information from Steps 4 and 5; the results section draws from Step 6; and the conclusion summarizes Steps 7 and 8. In these reports students strive to write without extra clutter. They capture the experiment in clear language so that another scientist could replicate their results. A student scientific report based on the density experiment appears in Figure 3.

In this class, the teacher continued modeling using the steps outlined in the laboratory guidelines until her students could follow the procedure on their own. She gave similar instruction for developing reports. The class did several reports together until they succeeded in writing their own. During this period of transition, she assigned group laboratory reports. Each lab group (three students) completed the experimental guide, conducted the experiment, and wrote a report. The teacher requested final copies on dittos, which she

Figure 2
Student Laboratory Notes

1. *Purpose.* To understand density and to determine the density of various objects, rocks in the earth's crust, and water.
2. *Problem.* We are going to determine how to calculate density and how it is used. Also, we want to see if the size and shape of an object affect its density.
3. *Hypothesis A:* I don't think size will affect the density, but I think shape might. If something is flat I think that will compress it and make it more dense.

   Hypothesis B: To me all rocks seem about the same heaviness, so I would think all the rocks in the crust would have close to the same density. In this experiment we are going to be comparing rocks to other materials such as glass spheres, aluminum cubes and bars, and steel spheres. I think the density of rocks would be about like the bar. The rocks are probably more dense than the cubes and glass and less dense than the metal ball.
4. *Materials.* Density kit: 1 aluminum bar, 2 aluminum cubes, 1 glass sphere, 1 steel sphere; 3 rocks; ruler; balance; cylinder.
5. *Procedure.* Hypothesis A: (1) Find the mass of each item in the density kit, using a balance; (2) find the volume by water displacement; (3) divide mass by volume; (4) record all information on a chart.
   Hypothesis B: (1) Find the mass of rocks A, B, and C, using a balance; (2) find the volume by water displacement; (3) divide mass by volume; (4) record all information on a chart.
6. *Data or Results.* Hypothesis A: Densities of items in the density kit.

Object	Mass	Volume	Density
Aluminum bar	34.2 g	12 cm^3	2.7 g/cm
Aluminum cube	5.4 g	2 cm^3	2.7 g/cm
Aluminum cube	5.4 g	2 cm^3	2.7 g/cm
Steel sphere	8.6 g	1.1 cm^3	7.8 g/cm
Glass sphere	2.9 g	1.1 cm^3	2.6 g/cm

   Hypothesis B: Densities of rocks from the earth's crust.

Object	Mass	Volume	Density
Rock A	40.1 g	14.8 cm^3	2.7 g/cm
Rock B	66.3 g	23.2 cm^3	2.9 g/cm
Rock C	18.4 g	7.3 cm^3	2.5 g/cm
Water	4.1 g	4.0 cm^3	1.0 g/cm

7. *Analysis or Conclusion.* The results from the first experiment indicate that the size and shape of an object do not affect its density. The bar and two cubes were made of aluminum. Even though the bar was larger and a different shape, it still had the same density. I can see that my hypothesis that the flat bar would be more dense is wrong, because the bar was just cut flat. If the bar were pressed, the length and width would stretch out (like clay), so the volume would probably still be the same (I think this might be a good lab to try). Also, even though the two spheres were the same size and shape they did not have the same density, so it must be the composition of the material that determines the density. In the second experiment, I found that the rocks varied from 2.5 to 2.9 g/cm^3. That seems close to the same value, as I predicted. The density of water is 1 g/cm^3.
8. *Class Conclusion.* The average of the class data was very similar to mine, so I believe my calculations were right. My conclusions were like most of the class's as well. Looking at all the data collected from part two of the lab, it looks like all the rocks in the earth's crust have densities that vary from 2.5 to 3 g/cm^3. I think probably each mineral, like aluminum, has its own density. Rocks are made up of different combinations of minerals, so their densities are different, but still close to the same value.

---

reproduced for the whole class to read and evaluate. From this evaluation the class developed a checklist of the features a scientific report should contain (Figure 4). This checklist provided the guidelines for writing and evaluating subsequent reports.

Each class should draft its own report checklist. The process of development is perhaps even more important than the final product. The discussion is invaluable because it gives students ownership over their evaluation instrument.

After writing reports in laboratory groups and evaluating them as a class, most students are ready to begin writing on their own. Before turning in their reports, students are first asked to read one another's drafts in reaction groups. The knowledge that peers will read the draft encourages neatness and clarity. In addition, when students read one another's work, they learn about both the content and the process of science.

We recommend developing reaction group rules to ensure that these occasions become useful learning experiences. Teachers usually divide students into groups of three. One person reads his or her work to the other two students, who listen and provide initial reactions. Next, they use the report checklist to analyze the draft. If some points are missing, the students provide suggestions for including these items. Once this process has been completed for one student, the cycle begins with the next person in the group. The students then revise their drafts before submitting them to the teacher for evaluation.

We recommend introducing the reaction group process with a demonstration. Two students can form a group with the teacher in front of the class. The teacher takes on the role of a student and reads a paper to the two reactors. They discuss how to provide feedback using the checklist. If students know how to offer support to one another, reaction groups

Figure 3
Student Scientific Report

1. *Purpose.* In this experiment we examined the densities of different objects to see if size and shape had an influence on the density. We thought the size would not affect density, but the shape might. We also calculated the densities of about 50 different rocks in the earth's crust to see if we could draw any conclusions. We thought the densities would be close in value, and about the same as the aluminum bar.
2. *Materials and Procedure.* To test whether size and shape had an effect on density we calculated the densities of two cubes and a bar composed of the same material, aluminum, but having different sizes and shapes. Next we calculated the densities of two spheres having the same size and shape, but composed of different materials, glass and steel. Finally, we found densities for two cubes having the same size and shape and composed of the same material, aluminum. In order to evaluate the density of a sampling of rocks from the crust, each lab group evaluated the density of three rocks. As a comparison, and because a lot of the earth's surface is covered with water, we also determined the density of water.
3. *Results.* We found that the aluminum objects, no matter what size or shape, always had the same density, 2.7 g/cm^3. Although the spheres were the same size and shape, the steel was a lot more dense (7.8 g/cm^3) than the glass (2.6 g/cm^3). In the second part, we found that the densities of the crustal rocks varied from 2.5 to 3.0 g/cm^3. We found that water has a density of 1.0 g/cm^3.
4. *Conclusions.* Our data indicated that size and shape do not affect the density of a certain material. We learned that the rocks in the earth's crust are 2.5 to 3.0 times as dense as water, since water has a density of 1. Water has a density of 1 because when the metric system was set up, water was used as the unit of measure.

are very effective. Developing guidelines and modeling how to use the guidelines are the keys to success.

Most teachers do not assign laboratory reports for each experiment. Students generally turn in at least four written reports per quarter. These reports have progressed through the process of drafting, evaluation by reaction groups, and revision.

In assigning scientific reports, our goal is not to teach students how to write but to teach the scientific process. Writing clears confusion in thinking. If students' reports are unclear, they haven't thought through the procedure. If they cannot write a logical hypothesis and then support or refute the hypothesis in the conclusion of the experiment, they have missed the main point of the experiment. Students become adept with the scientific method more quickly if they have to write about what they do. Clear writing and success at each step of experimentation are the same process.

Figure 4
Laboratory Report Checklist

1. *Purpose*
   - ☐ Have I explained why I am doing this experiment?
   - ☐ Did I conclude this section with a hypothesis?
2. *Materials and Procedure*
   - ☐ Did I explain the materials and procedure?
   - ☐ Did I explain the steps used to test my hypothesis?
3. *Results*
   - ☐ Did I present the data from the experiment?
   - ☐ Are the differences among variables clearly presented?
4. *Conclusions*
   - ☐ Are my conclusions directly based on the data?
   - ☐ Did I refer to the hypothesis?
   - ☐ If the data did not support my hypothesis, did I provide some reasons for the discrepancy?

Explanations of Scientific Phenomena

Asking students to take on the role of teacher and provide their own oral and written explanations for phenomena is another way to encourage active learning and thinking. None of us ever forget our first year of teaching, when we finally understood our content because we had to teach it. When research on learning is examined, it is little wonder that a teacher learns content by teaching. A host of research supports the importance of active learning. The more mental energy exerted, the better the learning. It is impossible to teach effectively and remain a passive learner.

We can create the same active integration for our students by having them teach one another. In fact, when content is particularly difficult, we institute the "drastic" strategy of involving students as teachers and writers. Simply stated, students work in pairs to explain concepts to one another and then put their explanations on paper. This strategy is an adaptation of ideas drawn from work on cooperative learning (Dansereau, 1985) and reciprocal teaching (Palincsar & Brown, 1986), which have records of resounding success.

After introducing the reading, the teacher assigns students to work in pairs. Both silently read the first few paragraphs or sections of the assignment. Then the "teacher" of the pair puts the material out of sight and explains the content to his or her partner. The partner asks questions. Next, both students refer back to the text to clarify the answers to the questions. Then the students proceed with the reading, but with roles reversed. The teaching and questioning cycle continues until the students finish reading the assignment. Each pair then drafts a written explanation of the content with their books closed. After completing their drafts, students can look back in their books for clarifying information before revising their work.

To encourage clear and complete explanations, teachers ask students to write for an audience that knows nothing about the content. When students write for the teacher, their explanations lack precision and clarity because they know the teacher already understands the content. Why should they write detailed explanations for a knowing audience? If the intended audience knows nothing, however, explanations take on a different dimension. Within these guidelines, we allow students to choose their own audience—for instance, a parent, a younger sibling, or friend.

We also encourage variations of roles and formats. Students may choose who they are as writers. They don't always have to be themselves; they can take on other roles, such as a blood cell, a pollen grain, or an amino acid. They also can vary from the traditional essay format, opting for letters, editorials, diaries, obituaries, or memos.

Writing experts tell us that students need to choose their own topics for writing (Calkins, 1987; Graves, 1983). School writing, they say, should mirror the work of real writers; it should be on topics of personal value for real audiences. As we think of ourselves as writers, we understand the validity of this argument. Yet science teachers want their students to write about important science content. Allowing students to choose their own roles, formats, and audiences when explaining a topic preserves some freedom of choice even when the topic is constrained. Explanations become livelier and more fun to read.

In one case, a teacher assigned students working in pairs to read a selection on the circulatory system. After reading half of the assignment, one student in each pair acted as teacher and explained the information while the partner listened and asked clarifying questions. Each pair switched roles for the remaining half of the selection, and then jointly wrote an explanation, choosing their own role, format, and audience.

One pair decided to write their explanation in a travelogue format. They took on the role of a red blood cell writing to a human audience that knew nothing about blood cells. Their explanation described the function of the cell and its travels through the circulatory system:

Student Writing Example

Hello human body. I am your very own red blood cell. I want to tell you a few things about myself, and then talk about how I travel throughout the highways in your body.

I am a cell in your blood. I carry oxygen. I am quite small. If you were to line me up in a row of 150 red blood cells, we would be only 1 millimeter long. That is about as wide as your pencil lines. I am round with thick edges and a thin center. I have 25 trillion red cell buddies all traveling around in your vessels. That's enough about me. Now on with my travels. I am like a delivery man. I help with the blood's job of being a pickup and delivery service, but I have lots of fun sliding around in your vessels. The oxygen gas coming into your lungs is picked up by me as I pass through the capillaries in the lungs. The lung is my favorite part of the trip. I splash around in the lungs while my hemoglobin picks up oxygen. I perform the job of pickup and delivery so well.

Now I am on my journey again to the heart. I am making the pulmonary trip from the lungs to the heart. Now I am inside the heart. It is very slippery riding up and down and around curves and out of valves. Finally, I leave the left ventricle and begin the systemic journey carrying my oxygen to the cells. The blood vessel is a wet-and-wild place to be. We slide down branches of the vessel, similar to climbing up the highest branch on a tree until we find just the right twig, which is like a capillary. Finally my buddies and I squeeze single file into these tiny capillaries where we unload our oxygen. We then get back together again in veins and work ourselves back to the heart. The heart pumps us through the pulmonary system to the lungs where our journey begins again. Off to deliver more oxygen to tired cells.

Conclusions

There was a time when science teachers in our district thought writing had no place in their classrooms. Writing across the curriculum was a buzzword of the administration and a few diehards in the English department. The science teachers had enough trouble just covering content; they certainly did not want the extra task of writing.

Recalling these thoughts, we realize how far we have come. Our science classrooms are far richer now. Learning logs have added vitality to learning. Students have the freedom to write anything they wish in their logs. They use these journals to capture their thinking on paper. Writing also empowers students to grasp the complexities of the scientific method. Their thinking travels from the problem to the conclusion, and as they write they internalize scientific patterns of thought. Finally, we know our students understand when they can teach difficult concepts through oral and written explanations.

As we examine our growth as teachers, our knowledge is similar to the bud of a flower. We know a little, but we are still emerging. Fortunately we have our classrooms for observing our students read, write, talk, and experiment. By continuing to take our nutrients from them, we may make it as sunflowers yet.

References

Calkins, L. (1987). *The art of teaching writing.* Portsmouth, NH: Heinemann.

Dansereau, D.F. (1985). Learning strategy research. In J.W. Segal, S.F. Chipman & R. Glasser (Eds.), *Thinking and learning skills, 1* (pp. 209-240). Hillsdale, NJ: Erlbaum.

Graves, D. (1983). *Writing: Teachers and children at work.* Portsmouth, NH: Heinemann.

Palincsar, A. S., & Brown, A.L. (1986). Interactive teaching to promote independent learning from text. *The Reading Teacher, 39,* 771-777.

CHAPTER 14

Teaching Vocabulary to Improve Science Learning

Bonnie C. Konopak

Konopak presents a host of practical approaches for teaching difficult science vocabulary. She offers instructional guidelines useful for both vocabulary instruction and teaching in general. Then, as she demonstrates each strategy with the same piece of science text, we see how graphic organizers, possible sentences, concept maps, list-group-label, and feature analysis can be used successfully at any grade level.

In the sciences, subject matter is distinguishable by the technical terms that label the important concepts. Acquisition of this vocabulary is important since students are expected to know and use these words for studying content material. As Herber (1978) states, "If students hold limited meanings for the words, they will hold limited understandings of the concepts, and hence limited understandings of the subject" (p. 130). Consequently, vocabulary development is an important component of learning in the science areas.

While science teachers recognize the importance of vocabulary learning, they may not know how students develop meanings. In addition, they may be unaware of recent instructional practices based on conceptual development. Given these considerations, this chapter will discuss the principles underlying content word learning and offer guidelines for instruction and specific teaching strategies that enhance vocabulary development and comprehension for science learning.

Types of Vocabulary

In a ninth grade general science class, the students may be assigned a selection such as the one presented in Text Passage 1.

TEXT PASSAGE 1

According to atomic theory, an atom consists of a small, dense nucleus surrounded by mostly empty space in which electrons move at high speeds. Most of an atom's volume is empty space. The average diameter of a nucleus is about 5×10^{-13} centimeters. The average diameter of an atom is about 2×10^{-8} centimeters. The difference in these two sizes means an atom is about 40,000 times larger than its nucleus. Consider an example of this relative difference. If the nucleus were the size of an orange, the whole atom would measure about 24 city blocks across.

Because atoms are a difficult concept, the textbook authors provide a familiar, concrete analogy to enhance students' understanding. However, the material still may be difficult to read. As Vacca and Vacca (1986) note, the words get in the way.

In the sciences, vocabulary consists of general and technical terms. General words are those with common meanings that are not associated with any one content area, such as *change* and *surface.* Technical words are those that apply to a particular subject. They may include general terms used in a special way, such as *matter* and *mass,* or terms that have only one distinct meaning and application, such as *photosynthesis* and *polyunsaturated.*

In the paragraph on atoms, general vocabulary words include small, dense, empty, difference, example, and orange. Technical words include atom, nucleus, electrons, space, diameter, and centimeters. Students probably will know and be able to use the general terminology, having encountered it many times in the past. However, they may have problems learning the technical words if they lack the related experiences and understandings from which meanings are derived. To understand the concept of atomic theory, these words must be understood on an individual basis and as interrelated parts—a difficult task without instruction.

Vocabulary Development

Herber (1978) distinguishes between having a definition for a word and having a meaning for it. A definition provides some facts about a word in isolation, which Herber refers to as "form without substance, a body without life" (p. 135). While a definition is a start, a meaning is not formed until the word is connected with other words within a particular unit of study. At that point, it begins to "develop its own character, its meaning for the user" (p. 135).

In the sciences, the word *development* aptly describes new vocabulary learning. Here, the emphasis is on gradually extending knowledge from a definition to an elaborated meaning. While teachers may assume that using a dictionary to define a word is sufficient for learning, research points to the multifaceted nature of vocabulary knowledge (Drum, 1983). Rather than an all-or-nothing proposition, vocabulary knowledge is a continuum, ranging from "complete unfamiliarity...to a trace of knowledge, to accurate but narrow knowledge, to fluent and rich knowledge" (Beck & McKeown, 1985, p. 11). For example, students may have a general understanding of an atom as "the smallest piece of matter." However, they may not know its composition, size, or relationship to larger units.

Development refers to those characteristics of learning that gradually lead to a rich understanding of a word: multiple exposures within a variety of contexts. Burmeister (1978) provides an analogy. As with words, we know people on a continuum. There are people we know only by name; there are others we have met on one occasion or several times in settings that provide us with limited information; and then there are those people we know very well, having met them over a period of time in many different situations. For example, we may not learn that a friend is allergic to chocolate until the day we serve chocolate mousse for dinner, or that she has an identical twin until we greet her look-alike on the street (Moore et al., 1986). To expand students' knowledge of atoms requires multiple experiences with the concept in different settings.

Conditions for Learning

Development is inherent in the notion that a word represents a concept or schema (Readence, Bean, & Baldwin, 1985). A schema is defined as a cognitive framework in which we store our past experiences and knowledge. Because we are always encountering new experiences and information, this framework is in a constant state of flux. Schema theory attempts to explain how we learn and how we cope with new experiences. Two alternative processes are suggested: assimilation of new information into an existing schema, and accommodation of radically new or discordant information through the modification or creation of a schema. Schemata develop and change with incoming information.

These processes are similar to two learning conditions classified by Graves (1987). The first condition is learning words that are not in the students' oral or reading vocabularies but for which they have an available concept. An example is the word *indigenous,* which may be unknown to students, although they understand the concept of native born. Here, a new label is being learned for an already-acquired concept. Such learning is assimilation, whereby new information (indigenous) is incorporated into an existing schema (native born).

The second and more complex learning condition occurs when no concept is available. One example is the word *force,* for which the students may have a general but not a technical meaning. A second example is the word *fulcrum,* for which the students may have neither the label nor the concept. This learning may be described as accommodation, since schemata must be modified or created to accept this new information.

For the most part, science teachers are concerned with helping students learn novel information and develop concepts. Because the focus is on accommodation, instruction is essential. We need to ensure that students have several encounters with new vocabulary terms in different learning situations to extend their knowledge from a definition to an elaborated meaning. In doing so, the students must be active learners and continue to develop schemata for new science concepts.

Guidelines for Vocabulary Instruction

The purpose of vocabulary instruction is to enhance students' understanding when they study content materials. Not all strategies meet this objective, however; while some studies have found that teaching word meanings enables better understanding, others have failed to find such an effect (Mezynski, 1983). Because not all procedures are equally effective, guidelines based on learning principles have been developed to help teachers evaluate vocabulary instruction (Carr & Wixson, 1986):

1. Instruction should help students relate new vocabulary to their background knowledge. Effective instruction helps students associate their past experiences with the new words. In this way, a bridge is provided between new and known information, which leads to increased comprehension.
2. Instruction should help students develop elaborated word knowledge. Effective instruction aids students in developing a meaning, not just a definition, for a new word in relation to its larger topic. Generally, this involves not only introducing new vocabulary but also providing reinforcement activities that help students review and apply their new knowledge.
3. Instruction should provide for active student involvement in learning new vocabulary. Effective instruction ensures that students are active, rather than passive, learners in developing word meanings. This means providing activities that focus on student-centered, rather than just teacher-directed, learning.
4. Instruction should develop students' strategies for acquiring new vocabulary independently. Effective instruction emphasizes the role of the student in becoming an independent word learner. Because of the tremendous number of words students need to understand, it is important for them to learn *how* to learn. While this task can be particularly difficult in the sciences,

Figure 1
Matrix for Evaluating Vocabulary Strategies

Method	Background Knowledge	Elaborated Word Knowledge	Active Student Involvement	Independent Learning
Strategy 1				
Strategy 2				
Strategy 3				

Adapted from E. Carr & K. Wixson, "Guidelines for Evaluating Vocabulary Instruction," *Journal of Reading,* April 1986, p. 592.

teachers can help by modeling appropriate word selection, providing independent strategies and practice time, and monitoring students' word meaning development.

To assess vocabulary strategies, Carr and Wixson (1986) offer an evaluation procedure that incorporates these four guidelines. Teachers can rate each strategy numerically in terms of its emphasis on students' background knowledge, development of elaborated word knowledge, active involvement, and independent learning (see Figure 1). Carr and Wixson stress that the guidelines should not be thought of as individual factors. Rather, "they represent an attempt to describe complex, interrelated processes in a way that is useful for educational practice" (p. 589).

The following sections describe instructional practices for determining key vocabulary to be studied, as well as methods of introducing and reinforcing new meanings. These practices support the four guidelines, emphasizing the need for students to be active and independent learners in developing conceptual knowledge.

An excerpt on the classification of matter, taken from a ninth grade general science textbook (Moyer & Bishop, 1986), is used to demonstrate these practices (see Text Passage 2). The passage contains a general description of heterogeneous and homogeneous matter, applicable to both life sciences and physical sciences. The text passage is used throughout the discussion as an example of how these techniques might be implemented.

TEXT PASSAGE 2

Classification of Matter

In order to avoid confusion, some method is needed to classify the millions of different types of matter. A classification system is an organized way of putting similar objects into groups. People make use of classification systems every day. Telephone books are a common example of a classification system. How are names and numbers classified in telephone books? Imagine trying to locate a phone number without this classification system.

Many classification systems for matter have been proposed. Aristotle (384-322 B.C.) suggested that all matter could be classified into four groups: earth, air, fire, and water. He thought that a fifth type of matter might exist. He called it the "perfect substance."

Today, scientists classify matter as homogeneous or heterogeneous. *Homogeneous* matter is the same throughout. *Heterogeneous* matter can be separated into two or more types of homogeneous matter.

All heterogeneous materials are *mixtures*. When you make a tossed salad at a salad bar, you make a heterogeneous mixture. You start with some lettuce on your plate. Maybe you like a lot of cheese on top and a few mushrooms. Your salad is not the same throughout. Each bite is slightly different. At first you taste a lot of cheese and an occasional mushroom. At the end you taste mostly lettuce. Heterogeneous matter is not the same throughout.

Some types of homogeneous matter are mixtures. Mixtures that are homogeneous are called *solutions*. Suppose you order a cup of tea with your salad. You may stir sweetener into the tea, making a solution of tea and sweetener. The tea tastes sweet when you start drinking it and it tastes sweet when you finish. The amount of sweetener is the same throughout the tea. Homogeneous matter is the same throughout.

Not all solutions are liquids. Deep-sea divers breathe a gaseous solution of oxygen and helium while they are under water. The alloy sterling silver is a solid solution of silver and copper. An alloy is a combination of two or more metals or a metal and one or more nonmetals. Alloys can be heterogeneous mixtures, solid solutions, or compounds.

Substances are another type of homogeneous matter. A *substance* is homogeneous matter that always has the same composition. Water is a homogeneous substance. The composition of water, H_2O, is always the same—two atoms of hydrogen and one atom of oxygen.

Substances are divided into two classes: elements and compounds. The simplest substances are elements. An *element* is composed of any one type of atom. Scientists have discovered 109 elements. Ninety elements occur naturally in the earth. The remaining elements have been produced in laboratories. Most of the elements are quite rare. Some of the most common elements are oxygen, silicon, aluminum, iron, and calcium.

When two or more different elements combine chemically, a *compound* is formed. Compounds are a more complex class of substances. Compounds are different from the elements from which they were formed. Rust is a compound that is composed of the elements iron and oxygen. The element iron is a metallic solid while oxygen is a colorless, odorless gas. The properties of these two elements are quite different from the properties of rust.

Vocabulary Selection

Science teachers cannot teach all of the unfamiliar words students might encounter in text. However, by analyzing the text and listing the key concepts to be learned, teachers can decide which terms the students need to understand. These "usable" words appear frequently in explaining the concepts under study and are key to understanding other words (Haggard, 1982).

Moore et al. (1986) offer two commonsense steps science teachers can follow to select relevant words: (1) Analyze the unit to be studied and list important unknown words pertinent to the topic, and (2) pare down the list by first selecting words that are important for understanding the content area in general and then adding words that appear repeatedly and are crucial to understanding the particular unit.

With her vocabulary self-collection strategy (VSS), Haggard (1982) suggests that teachers help their students select their own words for study. To do this, have students preview materials to be read and identify two words they consider important and relatively unknown. Point out that important words occur many times in the text or are highlighted in some manner. Then modify the class list to delete less important words or to add missing crucial words, showing the students how these modifying decisions were made.

To prepare for instruction, Stahl (1986) suggests that teachers try to decide how likely students are to get the meaning of the word from textbook context. If a key word is defined and elaborated on in the text, it may require less attention in class.

Once you have a list, decide whether each word is a new label/old concept type or an old-new label/new concept type. While the first type can be taught relatively easily by associating the new word with known words and concepts, the second type requires more extensive instruction.

In the passage on classification of matter, the following words may be considered key vocabulary: heterogeneous, homogeneous, mixtures, solutions, substances, elements, and compounds. These terms are important for understanding the topic of classification and provide a basis for further study of more complex concepts. In addition, each is highlighted in the text and occurs several times. For instructional purposes, each word is at least partially defined in context, with some examples included. Vocabulary instruction would be useful to help students better understand the terms (such as element and compound) that may introduce new concepts as well as to understand how these terms are related.

In the next sections, methods of introducing and reinforcing word meanings are described. Students need initial encounters with unfamiliar key words, along with additional opportunities to review and strengthen the associated meanings. Accordingly, teachers need to prepare introductory lessons, as well as activities for review and practice. In addition, students need to become independent learners through instruction that gradually shifts responsibility for learning from the teacher to them.

Introducing Word Meanings

Teaching new vocabulary before students read prepares them in two ways: it provides them with background knowledge necessary for understanding the topic, and it gives them direction and purpose for meaning. Having this information not only enhances students' comprehension of text materials, it also indicates the important information on which to focus. As Thelen (1984) notes, if students are forced to learn new material before they master the necessary backlog of experience for the concept, they may become frustrated and resort to memorizing definitions and trivia.

The following activities provide background information and focus on important information to promote word learning. The first activity involves a graphic organizer (Earle & Barron, 1973), a visual aid that defines the hierarchical nature of the concepts to be studied. The second activity, possible sentences (Moore & Arthur, 1981), helps students determine word meanings by pairing unknown with known words in sentences that could be encountered during reading. The third activity is a student-directed strategy that uses concept maps (Schwartz & Raphael, 1985) to provide a framework for the components of word meanings. For each activity, the purpose, a description, and several examples are given, in addition to recommendations for use.

Graphic Organizers

A graphic organizer presents students with an idea framework of important conceptual relationships between content vocabulary. Because science vocabulary can represent difficult concepts, the framework helps structure both the material and students' learning as reading progresses.

There are six steps to developing and presenting a graphic organizer: (1) concept selection, (2) diagram construction, (3) evaluation, (4) presentation, (5) follow up, and (6) independent practice.

1. *Concept selection.* The teacher begins by selecting the important concepts to be learned and identifying the vocabulary necessary for learning them. To prevent the organizer from becoming overly complex, the

teacher can use only higher order concepts for initial selection:

Classification of Matter

heterogeneous	substances
homogeneous	elements
mixtures	compounds
solutions	

2. *Diagram construction.* The teacher then arranges the selected terms in a tree diagram that reflects a three-tiered structure:

Classification of Matter

heterogeneous
homogeneous
heterogeneous mixtures
homogeneous mixtures
substances
solutions
compounds
elements

3. *Evaluation.* Once the diagram is created, it should be evaluated to ensure that it accurately conveys the concepts the teacher wishes to present. At this point, modifications can be made if the organizer is not clearly constructed.

4. *Organizer presentation.* The organizer can be presented on the chalkboard or overhead transparency, beginning with a general explanation of its purpose and how it is arranged. The teacher then helps the students explore the new concepts and explains unfamiliar vocabulary.

5. *Follow up.* The organizer is retained as the students progress in reading about the concepts. It can be used as a point of reference throughout the unit, with additional information being added when needed.

6. *Independent practice.* Once the students are familiar with the nature and construction of graphic organizers, the teacher guides them in creating and developing their own frameworks. By beginning with completed organizers and gradually reducing the number of vocabulary terms, the teacher allows the students to become responsible for their own study aids.

A graphic organizer is a tremendous aid to students who are unfamiliar with a difficult and complex concept, as it presents both vocabulary terms and their relationships. It can be used by the teacher to explain new word meanings or with a second activity in which students predict word meanings. In addition, a partial organizer to be completed by the students can help them become independent learners. By allowing active student participation, a graphic organizer incorporates the four guidelines for effective instruction.

Possible Sentences

The second strategy, possible sentences, is a combined vocabulary/prediction activity designed to enable students to determine the meanings and relationships of unfamiliar words in reading assignments. Students make predictions about the relationships between the unfamiliar words, read to verify the accuracy of their predictions, and use the text to evaluate and refine their initial ideas.

The possible-sentences strategy consists of the following steps: (1) listing key vocabulary, (2) eliciting sentences, (3) reading and verifying sentences, (4) evaluating sentences, (5) generating new sentences, and (6) providing independent practice.

1. *Listing key vocabulary.* As with the graphic organizer lesson, the teacher begins by listing key vocabulary words (heterogeneous, homogeneous, mixtures, solutions, substances, elements, compounds) on the chalkboard or

transparency and pronounces the words for the students.

2. *Eliciting sentences.* The teacher then asks students to use at least two words from the list and make a sentence they think might be found in the text. These sentences are recorded on the chalkboard or transparency, regardless of the accuracy of the information, and the key words are underlined. Students may use words already in previous sentences as long as a new context is created.

- Homogeneous matter has all the same elements in it.
- A mixture has several different substances in it.
- Compounds are elements put together.
- A solution is heterogeneous because it has different substances in it.

3. *Reading and verifying sentences.* Students then are asked to read the text and check the accuracy of their possible sentences.

4. *Evaluating sentences.* After reading, each sentence is evaluated for accuracy, with the text available as a reference. On the basis of the evaluation, sentences may be left intact, refined, or omitted. For example, the four sentences generated in Step 2 need to be refined to include additional or more accurate information.

5. *Generating new sentences.* After evaluating the original sentences, the teacher asks for additional sentences to further extend students' understanding of the meanings and relationships of the new words. These sentences are also checked for accuracy against the text. Students should record in their notebooks all final acceptable sentences.

6. *Providing independent practice.* Once students are familiar with this strategy, the teacher can provide them with practice time in selecting key vocabulary, making predictions, and verifying their guesses. Working as a whole class or in small groups is particularly beneficial at the beginning so students can see one another's activities; gradually the students will be able to study on their own.

With this technique, students are actively involved in their own learning. They use their experiential knowledge to think of possible associations between new vocabulary and then evaluate their accuracy based on the textbook information. While the teacher initially selects the key words and assists in monitoring the students' progress, gradually they can assume the responsibility of predicting and verifying.

Concept Maps

In contrast to the first two techniques, the third is primarily student-directed. A concept map provides students with a way of using context clues independently and determining whether they know what a word means. As Schwartz and Raphael (1985) point out, many students have only a vague notion of what constitutes a definition; this procedure helps them conceptualize the components of a definition and determine whether they can fill each component.

There are six steps to presenting this technique: (1) selecting key vocabulary, (2) providing textbook contexts, (3) presenting the concept map framework, (4) completing the map, (5) discussing results, and (6) providing independent practice.

1. *Selecting key vocabulary.* To teach this technique, the teacher once again begins by selecting target vocabulary words (heterogeneous, homogeneous, mixtures, solutions, substances, elements, compounds) and listing them on the chalkboard or transparency.

2. *Providing textbook contexts.* Using the textbook, the teacher helps students locate pertinent definitional information about each term. For the word *element*, the excerpt in Text Passage 3 might be chosen.

TEXT PASSAGE 3

The simplest substances are *elements*. An element is composed of only one type of atom. Scientists have discovered 109 elements. Ninety elements occur naturally in the earth. The remaining elements have been produced in laboratories. Most of the elements are quite rare. Some of the most common elements are oxygen, silicon, aluminum, and calcium.

3. *Presenting the concept map framework.* The teacher draws the following framework on the chalkboard or transparency and explains that each of the three components is necessary for a good understanding of the word.

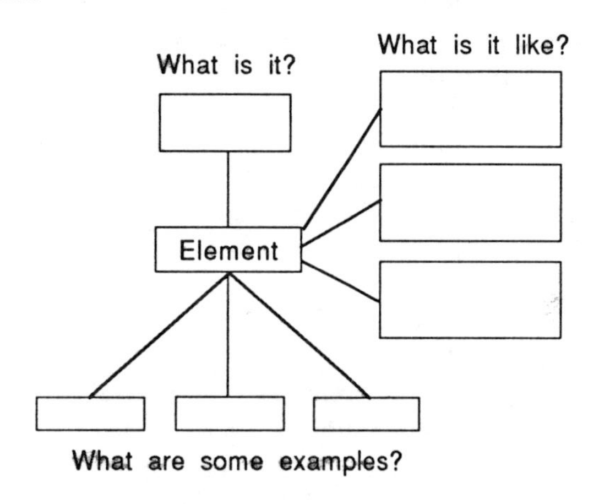

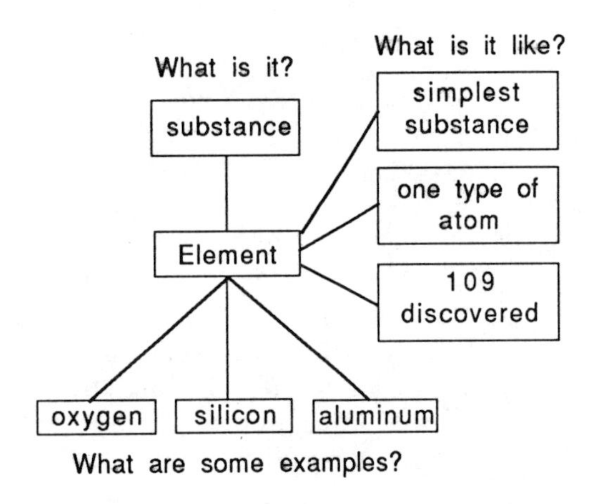

4. *Completing the map.* Using the textbook information, the teacher and the students complete as many of the framework components as possible.

5. *Discussing the results.* The teacher discusses the completed map with the students, emphasizing the three components. In addition, the teacher explains that not all contexts will provide these three types of information, so the students must monitor their own understanding.

6. *Providing independent practice.* Following the presentation and discussion of one or more example words, the teacher provides practice time for the remaining key vocabulary. After independent work, the class discusses the students' maps for the remaining words, providing feedback on their completeness and accuracy.

Using this strategy, students come to understand not only what constitutes a good definition but also how to evaluate the textbook explanation and their own comprehension. Once learned, this technique is particularly useful for key words that are well defined in context (allowing the teacher to focus instruction on other important vocabulary) and for words that are unfamiliar to the students but that the teacher may not include on the key vocabulary list for instruction.

Reinforcing Word Meanings

As Moore et al. (1986) note, prereading activities are good for introducing a word, but a good introduction is only the beginning. More encounters are needed in different contexts for students to review and reinforce the word meanings they have acquired. Again, these activities should focus on meaning—tying the word to the topic at large—as well as provide opportunities for students to use the word both receptively and expressively.

The following two activities are designed to provide for active student involvement in reviewing and using new words and meanings. The first activity, list-group-label (Taba, 1967), is a classification technique that emphasizes word relationships. The second technique, feature analysis (Johnson & Pearson, 1984; Pittelman et al., 1991), is a procedure to help students make fine discriminations among concepts. Again, the description of each activity is accompanied by an explanation of its purpose, some examples, and recommendations for use.

List-Group-Label

List-group-label is similar to the graphic organizer technique in that it stresses word relationships, but it is student-derived rather than teacher-provided. It actively engages students in reviewing important words and allows them to see how other students associate new words.

The list-group-label lesson consists of five steps: (1) topic selection, (2) listing procedure, (3) grouping/labeling procedure, (4) follow-up discussion, and (5) independent practice.

1. *Topic selection.* After the students have read the text material, the teacher selects a topic comprising important related terms (e.g., classification of matter).

2. *Listing procedure.* The teacher begins the lesson by selecting a one- or two-word term to serve as a stimulus for eliciting student responses (e.g., matter). The term is written on the chalkboard and students are asked to brainstorm words related to the topic. Student responses are recorded on the board, with all students being asked to generate an answer:

Matter

heterogeneous	alloy
homogeneous	mixture
109 elements	gas
compounds	rust
water	solution
salad	solid
oxygen	iron
atoms	liquid
substances	aluminum

3. *Grouping/labeling procedure.* The teacher begins this step by reading the list of words on the board and then instructing students to group these words into categories. These smaller groupings should consist of words that have something in common; one word may be used in several groupings. Finally, students are told to give each group of words a label or title that indicates the shared relationship they possess.

4. *Follow-up discussion.* After the students have finished grouping and labeling, the teacher records the categories on the chalkboard. After each category is recorded, the student offering the group then explains why the words were grouped together in that fashion. In this way, other students can see different category possibilities. Some examples follow:

Same Composition
homogeneous
substances
water
mixtures
alloy

Different Composition
heterogeneous
mixture
salad
alloy

Forms of matter
solid
gas
liquid

Compounds
elements
rust
oxygen
iron
alloy

Elements
atoms
109
oxygen
iron
aluminum

5. *Independent practice.* Following whole class/small group sessions, the teacher can guide the students into individually conducting a brainstorming and categorizing activity. By giving practice time, the teacher can encourage the students to use this technique as an independent learning activity after reading to activate and organize the new information.

List-group-label encourages students to explore newly acquired information in terms of common words and relationships. Individual students may complete the activity, or small groups of students may work together. In addition, teachers can find out what students have learned and what will require reteaching.

Feature Analysis

Feature analysis (or semantic feature analysis) is designed to help students improve their concept learning by understanding the similarities and differences among related words. Tierney, Readence, and Dishner (1990) state that as human beings process information, categories are established on the basis of cultural and experiential patterns. Rules are formulated to place words or concepts into these categories. In this way, category interrelationships are established so that individuals can search their category systems to make sense of their experiences. Here, feature analysis is intended to provide a systematic procedure for exploring and reinforcing vocabulary concepts.

Feature analysis consists of six steps: (1) selecting the category, (2) listing words in the category, (3) listing features, (4) indicating feature possession, (5) exploring the matrix, and (6) providing independent practice.

1. *Selecting the category.* The teacher selects a category that includes two or more similar items (e.g., classification of matter).

2. *Listing words in the category.* The teacher lists the category terms (heterogeneous, homogeneous, heterogeneous mixtures, homogeneous mixtures, substances, solutions) down the left side of the chalkboard.

3. *Listing features.* The teacher then records the features used to describe the category terms (same throughout, different throughout, evenly distributed, unevenly distributed, pure, separated physically) across the top of the chalkboard.

4. *Indicating feature possession.* The students then are guided through the matrix as they indicate whether each category item possesses a given feature. A plus (+) shows that the item usually has that feature; a minus (-) shows that the item usually does not have that feature. Students sometimes argue over a par-

Classification of Matter

	Features					
Terms	same throughout	different throughout	evenly distributed	unevenly distributed	pure	separated physically
heterogeneous	–	+	–	+	–	+
homogeneous	+	–	+	–	±	±
heterogeneous mixtures	–	+	–	+	–	+
homogeneous mixtures	+	–	+	–	–	+
substances	+	–	+	–	+	–
solutions	+	–	+	–	–	+

Note: + means yes; – means no; ± indicates that both conditions apply.

ticular item or feature; a majority of responses should decide the final plus or minus placement. (See completed matrix for classification of matter.)

5. *Exploring the matrix.* After determining feature possession, the students make observations based on the completed matrix. They can generate their own ideas, as well as respond to teacher-directed questions. Possible questions include differences between types of matter (mixtures and solutions) and differences between features (same and different compositions throughout). In addition, more complex questions can be asked, such as features of familiar types of matter (salad, rust).

6. *Providing independent practice.* Once the students understand the nature of the feature analysis task, the teacher can give the students practice time to add category items or features to an existing matrix and to practice creating their own matrices.

Feature analysis is an excellent technique for exploring the similarities and differences among related concepts. Students actively participate in completing the matrix by expressing their knowledge and, if challenged, providing a rationale for their decision. In addition, the strategy provides for independent responses as students extend original matrices and create new ones.

Conclusion

The five strategies presented all meet, in varying degrees, the four guidelines for vocabulary instruction offered by Carr and Wixson (1986). The students participate in their own learning: they activate and associate their prior knowledge with the new vocabulary, they develop elaborated word meanings and relationships, and they are actively involved in the learning process. In addition, through teacher guidance, these techniques allow students to gradually assume responsibility for their own learning.

In summary, because vocabulary knowledge is essential for science learning, direct instruction should be a major component of science classes. Technical words should be taught before reading to facilitate comprehension and reinforced after reading to confirm and extend their meanings. In addition, the instructional activities used should engage students in active learning through a variety of contexts, gradually leading them to become independent learners. Finally, as Readence, Bean, and Baldwin (1985) stress, the teacher should be an interesting and interested model of vocabulary use: ''Nothing will facilitate the acquisition of vocabulary more than the enthusiasm you convey to students'' (p. 96).

References

Beck, I., & McKeown, M. (1985). Teaching vocabulary: Making the instruction fit the goal. *Educational Perspectives, 23,* 11-15.

Burmeister, L. (1978). *Reading strategies for middle and secondary school teachers* (2nd ed.). Reading, MA: Addison-Wesley.

Carr, E., & Wixson, K. (1986). Guidelines for evaluating vocabulary instruction. *Journal of Reading, 29,* 588-595.

Drum, P. (1983). Vocabulary knowledge. In J.A. Niles & L.A. Harris (Eds.), *Searches for meaning in reading/language processes and instruction* (pp. 163-171). Rochester, NY: National Reading Conference.

Earle, R.A., & Barron, R.F. (1973). An approach for teaching vocabulary in content subjects. In H.L. Herber & R.F. Barron (Eds.), *Research in reading in the content areas: Second year report* (pp. 84-110). Syracuse, NY: Syracuse University, Reading and Language Arts Center.

Graves, M. (1987). The roles of instruction in fostering vocabulary development. In M. McKeown & M. Curtis (Eds.), *The nature of vocabulary acquisition* (pp. 165-184). Hillsdale, NJ: Erlbaum.

Haggard, M.R. (1982). The vocabulary self-collection strategy: An active approach to word learning. *Journal of Reading, 26,* 203-207.

Herber, H.L. (1978). *Teaching reading in content areas* (2nd ed.). Englewood Cliffs, NJ: Prentice Hall.

Johnson, D., & Pearson, P.D. (1984). *Teaching reading vocabulary* (2nd ed.). Orlando, FL: Holt, Rinehart & Winston.

Mezynski, K. (1983). Issues concerning the acquisition of knowledge: Effects of vocabulary training on reading comprehension. *Review of Educational Research, 53,* 253-279.

Moore, D.W., & Arthur, S. (1981). Possible sentences. In E.K. Dishner, T.W. Bean, & J.E. Readence (Eds.), *Reading in the content areas: Improving classroom instruction* (pp. 138-143). Dubuque, IA: Kendall/Hunt.

Moore, D.W., Moore, S.A., Cunningham, P., & Cunningham, J. (1986). *Developing readers and writers in the content areas.* White Plains, NY: Longman.

Moyer, R.H., & Bishop, J.E. (1986). *General science.* Columbus, OH: Charles E. Merrill.

Pittelman, S.D., Heimlich, J.E., Berglund, R.L., & French, M.P. (1991). *Semantic feature analysis: Classroom applications.* Newark, DE: International Reading Association.

Readence, J.E., Bean, T.W., & Baldwin, R.S. (1985). *Content area reading: An integrated approach* (2nd ed.). Dubuque, IA: Kendall/Hunt.

Schwartz, R., & Raphael, T. (1985). Concept of definition: A key to improving students' vocabulary. *The Reading Teacher, 39,* 198-205.

Stahl, S. (1986). Three principles of effective vocabulary instruction. *Journal of Reading, 29,* 662-671.

Taba, H. (1967). *Teacher's handbook for elementary social studies.* Reading, MA: Addison-Wesley.

Thelen, J.N. (1984). *Improving reading in science* (2nd ed.). Newark, DE: International Reading Association.

Tierney, R.J., Readence, J.E., & Dishner, E.K. (1990). *Reading strategies and practices: A compendium* (3rd ed.). Boston, MA: Allyn & Bacon.

Vacca, R.T., & Vacca, J.A. (1986). *Content area reading* (2nd ed.). Boston, MA: Little, Brown.

CHAPTER 15 *Using Guided Imagery to Teach Science Concepts*

Barbara J. Walker
Paul T. Wilson

Walker and Wilson describe how students can use mental images to visualize and understand science phenomena. They explain why imagery should be included in both elementary and secondary classrooms to help students understand and remember important science concepts. Then they take us step by step through a lesson on evaporation, followed by suggestions for helping students use imagery on their own. Their analysis merges imagery research in psychology with classroom practice in science.

Students reading science text often are confused by the volume of new vocabulary words and scientific facts that seem unrelated to their everyday experiences. They often read their textbooks without thinking about how the information fits together or how it fits with what they know. In addition, teachers get caught up in delivering the content of the science lesson without realizing what the students are (or are not) learning. Teachers need to assist students in applying what they already know to material they are learning and in elaborating on the key relationships in the scientific explanations (Anderson & Smith, 1984).

To understand and remember scientific concepts, children must relate them to what they already know about the world. When reading science textbooks, young children must use their own knowledge to understand and elaborate on the author's message. For example, suppose students read the excerpt in Text Passage 1, from a fourth grade science text.

TEXT PASSAGE 1

One molecule of water is so tiny that we cannot see it. However, if a crowd of water molecules gathers, we can see the crowd. The crowd of molecules forms a drop of water (Cooper et al., 1985, p. 164).

To understand this paragraph, fourth graders could use imagery and knowledge about crowds they have developed from previous experiences—for instance, crowds of people in the supermarket or at the county fair. They also may think about the size of raindrops, or of drops of water that splash into the sink when they wash their hands. From these experiences, they begin to understand the size of molecules: molecules must be very small if a crowd of them fits into a small drop of water.

The more students elaborate on their reading, using text information and prior experiences, the more they will remember. But strategies for remembering and elaborating seldom are taught in science classrooms. In fact, one study found that only 17 percent of high school students mentioned receiving instruction that covered strategies for learning from text (Suzuki, 1985).

Guided imagery can help students elaborate on information while reading science texts. Teachers can design science lessons using guided imagery to show students how to read strategically by focusing on what they already know and by elaborating on the key relationships underlying the scientific concepts.

Imaging and Strategic Processing

We access much of what we know, and make inferences about new experiences, through memory images. Memory images are images of daily life that accompany recall of events (Whitmer & Young, 1985). Each memory image is a sensory or perceptual representation of a genuine experience or combination of experiences. Most prior experiences are represented in images of specific events. Thus, they have visual-spatial characteristics in addition to their sensory-perceptual (physical) qualities.

Think about the water molecule example. The text prompts students to form mental images of a crowd and of a drop of water. The memory images that form in the students' minds derive from particular experiences, like being in a crowd or washing their hands. One student's imaginal coding of *crowd* might include the memory of being jostled in a throng of people at the county fair, as well as the less physically oriented memory of experiencing the closeness of the crowd.

When reading the water molecule text, students must use images to form relationships between their own experiences and the content of the text (McNeil, 1987; Sadoski, 1984). It is not sufficient for them to imagine a crowd of people being rained on. Instead, they must manufacture an unusual relationship—such as the crowd fitting inside the drop of water—in order to form the concept the text communicates. Their ability to engage in this strategic process will determine the quality of their comprehension.

People experience images in three ways: spontaneously, by prompting, and through self-direction. Spontaneous images occur to most of us without conscious effort during daydreams or flashes of insight. Solutions to problems and creative inspiration often result from spontaneous imaging. Prompted images come to mind as a result of key words encountered during listening or reading. When people are asked to think about Christmas, the images they experience are a result of the prompt. Self-directed imaging is a person's deliberate choosing of the image conjured up. For example, we might call to mind an image of someone we miss. Or with reading and learning, a student might use imaging to rehearse the phases of the water cycle in preparation for a test.

Judging from their fantasy play, many young children engage in spontaneous imaging. They also may be able to recall images in response to the prompting of key words in a text. However, if the teacher further prompts students by discussing how to relate the images to form concepts in the text, their comprehension will improve (Gambrell, 1981; Gambrell & Bates, 1986; Linden & Wittrock, 1981; Pressley, 1976).

Imagery Instruction

Guided imagery is a teaching technique that combines spontaneous and prompted imaging. We use a unit on the water cycle to illustrate how guided imagery can be used to teach science concepts.

Designing the Imagery

Guided imagery involves careful preparation. We began planning our lesson by reading the text several times to identify key relationships in the scientific explanation. For example, in the water molecule text we found two important concepts: (1) the water molecules in the three forms of water (gas, liquid, solid) differ in terms of the distance between the molecules; and (2) evaporation (increasing the distance between the molecules) and condensation (decreasing the distance between the molecules) result from adding heat energy to or subtracting heat energy from the molecules.

Next we selected an analogy from the text to use in the guided imagery procedure. Here is the text's explanation of evaporation:

> A drop of water is a great crowd of water molecules. Molecules are always moving. As they move, some of the water molecules leave the crowd (Cooper et al., 1985, p. 164).

Taking the scientific explanation of evaporation (when heated, molecules begin moving faster until they spin apart and become water vapor) and the text-based image of a crowd of molecules, we elaborated on the analogy so that in the guided imagery the relationship between increased heat and water evaporation would be clearly explained. Our thinking proceeded along these lines. We needed an image of a crowd where there was constant movement and where, if movement became faster, the image itself would disappear. We chose dancing. Dancing molecules provided a concrete image of a physical action. Also, the activity of dancing had properties that could easily represent the key scientific relationships among speed of movement, heat, and distance apart.

Finally, we addressed these scientific relationships. In our image, the sun came out and the molecules danced faster. As they danced faster and faster, they got warmer and warmer, which caused them to spin apart and leave the crowd. As more molecules danced away from the crowd, the water drop (the crowd of molecules) disappeared. Our images illustrated the concepts and explained the relationship between heat energy and evaporation.

The two basic steps of designing imagery lessons are (1) identifying the key scientific concepts, and (2) developing an analogy that depicts these concepts. It is best to start with an analogy already in the text and to develop a concrete image familiar to the students. The analogy should represent all the key aspects of the scientific concepts.

The Guided Imagery Script

If using guided imagery for the first time, it is helpful to develop a mental or written script. We constructed the following guided imagery script for a fourth grade class learning about the water cycle:

Relax....Close your eyes and listen to the sounds of this room....Change the sounds of the room to the sounds of a pond on a sunny day....You are sitting beside the pond and gazing at the water....The waves of water move back and forth, showing you about its life....You see that the water is made up of tiny molecules. Imagine the molecules dancing in the water as if they were on a crowded dance floor. This crowd of molecules has formed a water droplet. Watch them dance. They are dancing and dancing. As the sun warms them, they dance faster, getting warmer and warmer. The sun's rays keep using their energy to heat up the molecules. One of the water molecules is dancing so fast and gets so warm that it spins apart from the crowd of molecules on the dance floor....Leaving the crowd of dancing molecules, it evaporates and disappears into the air. One by one, the molecules get warmer and move faster and evaporate....As all the molecules get warmer and evaporate, the water droplet disappears....When all of the water molecules have disappeared...think about how this room looks again....Open your eyes.

Guided imagery scripts include several characteristics:

1. Use multisensory statements that take the students from the classroom to the setting of the guided imagery. Including a variety of senses increases the vividness and ease of imagery production (Galyean, 1985). To evoke different senses such as hearing, our script began by asking the students to close their eyes and listen to the sounds of the room.
2. Repeat words or phrases and use intermittent pauses, which allow time for the students to form and elaborate on their images. In this script, key words like *dancing, warmer,* and *faster* and phrases like "get warmer and evaporate" were repeated. Pauses were indicated by ellipses.
3. Include a separation statement at the end of the guided imagery that prompts the students to return to the classroom and open their eyes. The separation statement naturally leads into discussion about the topic.

The Guided Imagery Procedure

There are six steps in using guided imagery effectively. First, before beginning the guided imagery, the teacher explains why imaging can be helpful when learning complex scientific information: "When we read, we not only read words and think about their definitions, but we also see pictures in our minds of the events being described. When reading science, people often create pictures about the ideas to help them understand the scientific explanations. In fact, Albert Einstein imagined riding a ray of light when he made one of his discoveries. Like the famous scientists, you can use pictures in your mind to help you understand and remember complex scientific information."

Second, the teacher explains that images will be easier to create when the students relax both their minds and their bodies. In a calm, serene voice, the teacher says: "Today we are going to make pictures in our minds about the lesson before we read. First, I want you to relax in your chair. Make sure you are sitting comfortably."

Third, when the students seem relaxed, the teacher presents a guided imagery script: "You are sitting beside a pond and gazing at the water. The waves of water move back and forth showing you about its life. You see that the water is made up of tiny molecules."

Fourth, the teacher leads a discussion about how images can help us understand scientific information. Initially, students tell a partner about the pictures they made. All have an opportunity to represent their images in words. After a few moments the teacher explains, "The pictures you made when your eyes were closed can be expressed in words. These same pictures can help you remember what you read." The teacher talks about her own images, and explains how they helped her remember the information. The teacher concludes by directing students to complete the statement, "When my eyes were closed, I pictured...." Here the students make a few notes about their images.

Fifth, the teacher prompts the students to use their images while reading: "As you read the text, think of your pictures and compare your pictures to the author's meaning. Ask yourself, 'How can my pictures help me understand and remember the information?'" The students' silent reading of the text is followed by a discussion about the text content in relation to their mental pictures. The students compare their images to those in the text, and check their understanding. Then they talk about the important ideas and facts of the selection.

Sixth, the teacher asks the students to close their eyes and recall images of the key re-

Figure 1
Developing Self-Directed Imaging

1. Use a section of science text as a springboard for developing imagery scripts as part of whole class discussions.
2. Encourage groups or individual students to develop their own scripts and present them to the class.
3. Ask students to explain their images and the congruence between their images and the content of the text. Prompt students with questions leading to the elaboration and revision of images.
4. Gradually relinquish your modeling and prompting role as the students take increasing responsibility for their own imaging.

Students will find the following steps helpful for creating their own images (adapted from Clark et al., 1984):

1. Read a paragraph or chapter section.
2. Make a mind picture using key words and analogies.
3. Talk to yourself about your image.
   - If you cannot make an image, explain why, and go on to the next paragraph or section.
   - If you can make an image, compare it with the image you had from the previous paragraph or section of your book.
   - You may need to revise your mind picture by adding to it or subtracting from it.
   - If it is a new mind picture, describe it.
4. Check your image with the text to see if it matches what the text says. If necessary, revise your mind picture by adding or subtracting based on the information in the text.
5. Continue to the next section and begin again at step 1.

Teach this sequence for self-directed imaging after students have succeeded with the more teacher-directed guided imaging. After students practice the routine silently, lead discussions in which students talk about their images, their revisions, and the effectiveness of the entire process. Using student-directed imagery according to this plan gives students study routines for using their background knowledge to elaborate on and remember scientific concepts.

lationships, prompting them quite specifically: "Water molecules are always moving; when they get warmer, they spin apart and evaporate." This step reaffirms the congruence between the students' images and the concepts in the text and primes the students' review strategy. The students are instructed to review by saying to themselves, "When I want to remember the lesson about evaporation, I will think __________ and remember __________."

These six steps represent what might be a teacher's first try at using guided imagery. Although the teacher may choose to continue directing the guided imagery activity, we recommend that students eventually begin taking more initiative in directing their own imaging. The teacher would gradually release responsibility for the imaging activity to students. The information in Figure 1 describes how to turn the responsibility for imagery over to the students.

A Caution about Textbook Analogies

Science textbooks use simple analogies to help children understand scientific concepts. Most analogies require students to visualize common experiences in order to understand complex scientific relationships. Imaging the analogy can help students verify whether their reading is making sense (Gambrell & Bates, 1986). But an overly simple analogy can impede students' in-depth understanding of complex relationships. We found an example in the same fourth grade text we used for our original example (See Text Passage 2).

TEXT PASSAGE 2

You might think of the air as a sponge. A sponge holds water until it is squeezed. Cooled air is like a sponge being squeezed. As air is cooled, its molecules lose energy and slow down. They come closer together. Some of the water molecules in the air are squeezed out (Cooper et al., 1985, p. 169).

In this example, the initial analogy of the sponge holding an unseen quantity of water that is later released applies to the scientific concept, and assists somewhat in its explanation. However, the analogy can mislead students into thinking that for a raindrop to appear, the air or cloud must be squeezed. This is inconsistent with the text's scientific explanation of why the water molecules move closer together.

Furthermore, the analogy confuses the relationship between energy and condensation. Students may think the squeezing concept means that energy must somehow be applied to the air or cloud in order for condensation to occur, or that the molecules are actually pushed together. In fact, the molecules slow down and come together as liquid condenses because they lose energy.

Many textbook analogies create powerful images that successfully illustrate one aspect of a scientific concept but that neglect or obscure other aspects, and so fail to communicate the complete scientific explanation. Such analogies can mislead students, as Spiro et al. (1989) have shown in their research on the acquisition of medical concepts. For this reason, teachers need to guide students carefully in checking incongruencies both in the text and in their interpretations of scientific information.

When students in the class we studied read the text, the teacher questioned the congruence between the image of the dancing molecules and the sponge analogy. The students discussed the fit between the analogy and the images they were creating of the water cycle, and then figured out how the analogy did not apply to all aspects of the scientific explanation. The teacher was able to prompt them to revise and elaborate on their knowledge of condensation by using the more accurate image that had been developed with the dancing molecules.

When a science textbook's analogy does not adequately represent a scientific explanation, the teacher should try to elaborate on the appropriate images in the text or design new images. The teacher should accentuate the features of these images that do illustrate the key relationships in the explanation. The students then can be prompted to check the congruence between the correct images and the analogy in the text. In this context, the presence of inadequate analogies may be beneficial because students can learn more directly to be questioning, aggressive readers.

Conclusions

Guided imagery instruction is an effective tool for showing students how to use their prior experiences to elaborate on and check their understanding of science texts. Using familiar images, teachers can prompt students to elaborate on textbook analogies and thus improve students' understanding of unfamiliar scientific explanations.

Constructing a guided imagery experience focuses attention on what students already know, as well as on the key scientific explanations to be taught. During the lesson, the process of comparing the images with the text gives the teacher clear information about what the students are thinking and learning. Misconceptions can be discussed and revised as students form a more complete image of the scientific concept. Images then become part of the students' background knowledge, and can be integrated with further learning in the science unit.

Finally, teachers must gradually release control of the guided imagery procedure by helping students form and revise mental images on their own. When students become self-directed in their imaging, they can independently use images to elaborate science concepts while reading—and, as a result, increase their retention of the information.

References

Anderson, C., & Smith, E. (1984). Children's preconceptions and content area textbooks. In G. Duffy, L. Roehler, & J. Mason (Eds.), *Comprehension instruction: Perspective and suggestions.* White Plains, NY: Longman.

Clark, F., Deshler, D., Schumaker, J., Alley, G., & Warner, M. (1984). Visual imagery and self-questioning: Strategies to improve comprehension of written material. *Journal of Learning Disabilities, 17,* 145-149.

Cooper, E., Blackwood, P., Bolschen, J., Giddings, M., & Carin, A. (1985). *Science,* level 4 (pp. 162-171). Orlando, FL: Harcourt Brace Jovanovich.

Galyean, B. (1985). Guided imagery in education. In A. Sheikh & K.S. Sheikh (Eds.), *Imagery in education* (pp. 161-178). Farmingdale, NY: Baywood.

Gambrell, L. (1981). Induced mental imagery and the text prediction performance of first and third graders. In J.A. Niles & L.A. Harris (Eds.), *New inquiries in reading research and instruction* (pp. 131-135). Rochester, NY: National Reading Conference.

Gambrell, L., & Bates, R. (1986). Mental imagery of fourth and fifth grade poor readers. *Reading Research Quarterly, 21,* 454-465.

Linden, M., & Wittrock, M. (1981). The teaching of reading comprehension according to the model of generative learning. *Reading Research Quarterly, 16,* 44-57.

McNeil, J. (1987). *Reading comprehension: New directions for classroom and practice.* Glenview, IL: Scott, Foresman.

Pressley, M. (1976). Mental imagery helps eight-year-olds remember what they read. *Journal of Educational Psychology, 68,* 355-359.

Sadoski, M. (1984). Text structure, imagery, and affect in the recall of a story. In J.A. Niles & L.A. Harris (Eds.), *Changing perspectives on research in reading/language processing and instruction* (pp. 48-53). Rochester, NY: National Reading Conference.

Spiro, R., Feltovich, P., Coulson, R., & Anderson, D. (1989). Multiple analogies for complex concepts: Antidotes for analogy-induced misconceptions in advanced knowledge acquisition. In S. Vosniadou & A. Ortony (Eds.), *Similarity and analogical reasoning.* Cambridge, U.K.: Cambridge University Press.

Suzuki, N. (1985). Imagery research with children: Implications for education. In A. Sheikh & K.S. Sheikh (Eds.), *Imagery in education* (pp. 170-198). Farmingdale, NY: Baywood.

Whitmer, J.M., & Young, M.E. (1985). The silent partner: Uses of imagery in counseling. *Journal of Counseling and Development, 64,* 187-190.

PART 5

A Reflection on This Book

CHAPTER 16 An Evolution of Learning

John T. Guthrie

This book is about the acquisition of reading comprehension and scientific thinking skills in science education. Various specialists from reading research and science education have brought the tools of their training and the directives of their discipline to this volume. I began my journey through this book by seeking a synthesis of the editors' thoughts, a pattern for the teaching practices assembled, and a group of characteristics common to students in science education.

At first blush, learning from a textbook (reading skills) and scientific thinking (science process skills) seem highly unified. Reading is regarded by scholars as a series of mental processes by which the mind gains new knowledge or experience from printed language. Amid these phases we see a person reading and learning from a beloved source—the book.

In parallel to the definition of reading, scientific thinking may be construed as a series of mental processes by which the mind gains new knowledge from the natural world (Kuhn, Amsel, & O'Loughlin, 1988). It is astonishing how closely aligned these definitions seem. Only the final phases diverge. Printed language is the predicate for reading and the natural world is the object of scientific thinking.

Reading is a series of mental processes by which the mind gains new knowledge or experience *from printed language. Scientific thinking* is a series of mental processes by which the mind gains new knowledge or experience *from the natural world.*

Several authors in this volume suggest that these two domains are conceptually unified. Baker proposes that the process skills of science include observing, classifying, comparing, measuring, describing, organizing information, predicting, inferring, formulating hypotheses, interpreting data, communicating, experimenting, and drawing conclusions. She argues that the self-correction, error detection, organizational structuring, and other text-based processes intrinsic to metacognition are related to the science process skills. Champagne and Klopfer suggest that elaboration is needed for activating prior knowledge and for encouraging the social interaction that undergirds learning. They confess, however, that there is no empirical evidence to demonstrate the validity of their conjecture.

Padilla, Muth, and Padilla reinforce the position that science process and reading are unified. They contend that "the integrated science process skills of carrying out an experiment, interpreting the results, and drawing conclusions are analogous to reading a text, making inferences, and drawing conclusions." In addition, Alvermann and Hinchman suggest that the process approach to education includes observing, predicting, experimenting, and interpreting. They propose that both learning to read science textbooks and learning to think scientifically rely on these processes. In their view, the book and the natural world are alternative information sources that students use to acquire these common cognitive skills.

Does this unification approach work? Do you believe that reading processes are like the process skills of a scientist?

Muse for a moment about the activities of sailing a boat and riding a bike. Both are sports. Both require balancing, navigating, changing direction, increasing speed, decreasing speed, adjusting to shifts in the weight of other riders, and adapting to wind speed. This is a reputable number of common components. Yet no one proposes that biking and sailing require the same skills. These abilities are rarely found in the same people, and a good teacher would approach them quite differently.

Sailing and biking are not analogous. Although the same terms have been used to describe both forms of recreation, these terms are put into operation in different ways. The way one balances on a bike is different from the way one balances in a sailboat. The methods of changing direction have no connection.

Just so, the process skills of science and the cognitive skills of reading are highly distinguishable. They are characterized by their uniqueness rather than by their similarity. Allowing for simplification for the sake of brevity, Popper (1959), Kuhn, Amsel, and O'Loughlin (1988), and Sternberg (1986) agree that the deductive process skills of science (i.e., scientific thinking) consist of: (1) rationale construction, (2) hypothesis formation, (3) observation/data collection, and (4) conclusion and inference. For this discussion, textbook reading processes (Anderson & Pearson, 1984; Pressley & Ghatala, 1990) consist of: (1) prior knowledge activation, (2) purpose setting, (3) text encoding/vocabulary, and (4) interpretation and integration.

Highlights of the distinctions between these sets of processes follow. The rationale stage of scientific thinking is often a unique, first-time experience for the learner, whereas prior knowledge activation in reading comprehension involves recollecting familiar ideas. A hypothesis formed in scientific thinking specifies a state of the natural world (or of the phenomena of interest) that will exist under given conditions. In contrast, purpose setting during reading broadly describes a need of the reader. The observations in scientific thinking are designed by the learner and generated to test the hypothesis. By comparison, reading in text comprehension consists of recognizing words or *re*constructing the author's ideas. (The emphasis is on *re*.) These ideas were originally thought of and constructed by the author.

Forming a conclusion in scientific thinking does not entail creativity. It is simply a connection of evidence to expectation (i.e., hypothesis). In reading, by comparison, the act of interpretation is a unique synthesis. It merges new ideas with old knowledge in the mind of the perceiver.

The evolution of learning is responsible for the divergence of reading comprehension and scientific thinking. This process began when humans diverged from the other primates. Learning distinguished us from other primates and enabled us to develop language and tool making (Bronowski, 1973). In the first stage of evolution, brought on by language competence, learning divided into two types: experiential and vicarious (see Figure 1).

Vicarious learning then subdivided into the oral tradition and reading, as a consequence of writing, which appeared in the Eastern Mediterranean in about 5000 B.C. The oral tradition provided person-to-person conveyance of culture, whereas writing provided book-to-person learning opportunities. Reading further subdivided into literature and science during the Renaissance, when books became available in these areas (Graff, 1987). By the late nineteenth century, science divided into social/behavioral and physical/life sciences. Contemporary science writing may be divided into original works in the form of articles or single-authored books and the ubiquitous textbook.

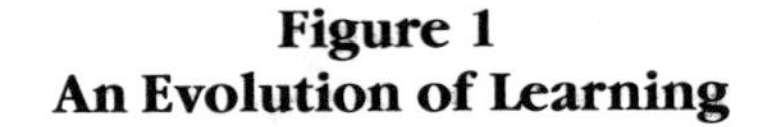

Figure 1
An Evolution of Learning

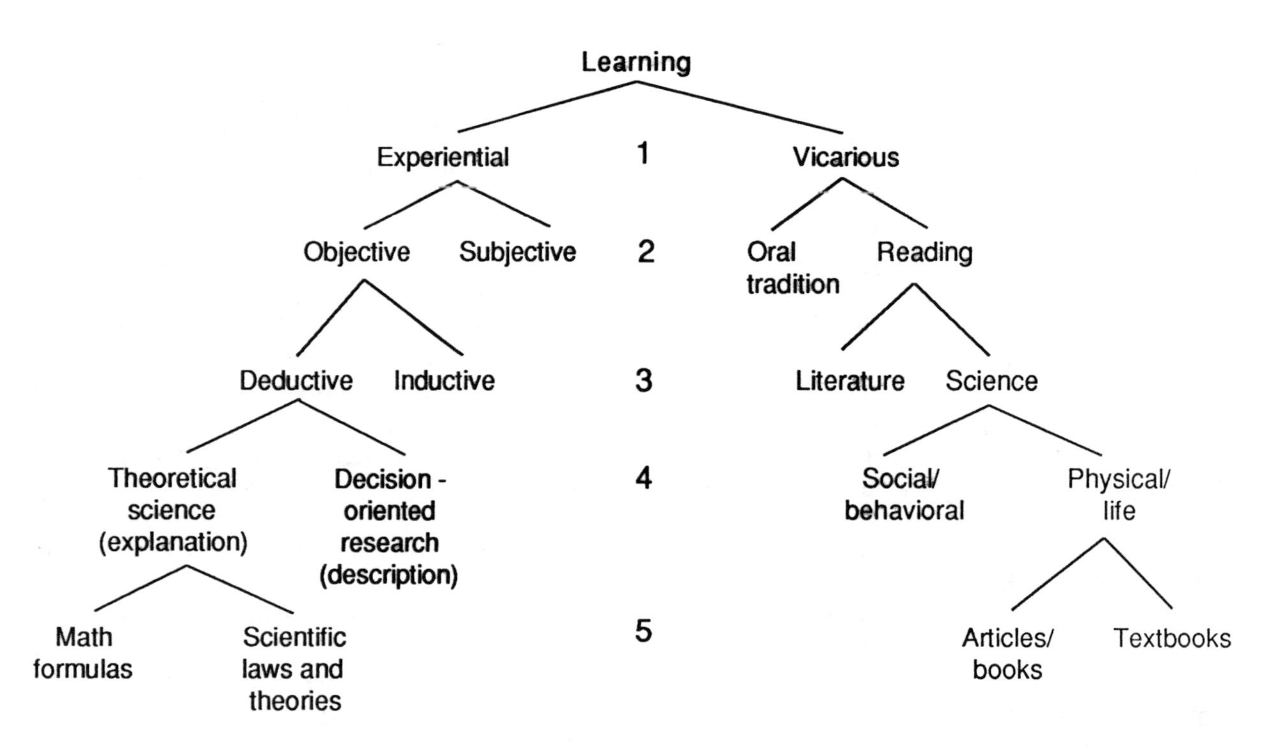

Experiential learning subdivided into objective and subjective when the scientific method provided an alternative to religious and philosophical ways of knowing. Within objective learning, the deductive-inductive contrast was articulated by Francis Bacon, who laid the basis of theory development and testing. Although induction was vital to Darwin, most current science is deductive (Foucault, 1972). Therefore, I have subdivided deductive scholarship into theoretical science and decision-oriented research. The former emphasizes explanation, while the latter is confined to factual description.

On the evolutionary learning tree, scientific thinking and reading comprehension are removed from one another. In their usual forms, learning science process skills and learning the cognitive strategies for reading comprehension are characterized by different critical features. Scientific thinking tends to be experiential, objective, deductive, and methodological. Comprehension of textbooks, however, is necessarily vicarious (mediated), print-based, expository, and conceptually demanding.

Many chapters in this volume are driven by the distinctions depicted in the evolutionary learning tree. For example, Alvermann and Hinchman portray one excellent teacher who avoids books but conducts animated science classes. He eschews the passivity of reading and believes that science process learning cannot be vicarious. In contrast to oral communication, reading requires metacognition. Baker, as well as Walker and Wilson, show that readers must detect their own errors of understanding, repair misconceptions during reading, and engage in other metacognitive activity. Because reading is solitary, and the teacher cannot always detect learning failures, the learner must assume this responsibility. Padak and Davidson advocate a cycle of predict/read/evaluate-the-prediction intended to place reading in a sociocommunicative context. This cycle endows reading with some of the personal immediacy of the oral tradition. Metacognitive strategies are an accommodation to the literacy base for vicarious learning, which is the first division on the text side of the evolutionary learning tree.

Science text and literary text differ in several dimensions. This distinction is the second split in the text branch of the evolutionary learning tree. According to Konopak, the uniqueness of science texts lies in their technical vocabulary. Holliday concurs by documenting the vocabulary burden of science texts. He proposes to reduce the number of terms taught at many levels of education, whereas Konopak proposes elaborate vocabulary teaching. Meyer contends that inadequate diagrams and charts pervade science texts; and textual confusion is notorious. These shortcomings flow from two circumstances: science often is impossible to represent in terms of daily experiences, and educators aspire to quickly cover vast terrains at all grades.

In high school, science means physical and life sciences. Although social and behavioral sciences have a secure footing in the culture, they are rarely taught in elementary or secondary schools, a distinction noted in the fourth stage of learning evolution. Student experience, being largely social, provides little bedrock for instruction. As Finley shows, incompatible background knowledge makes concepts difficult to learn. Roth proposes that direct instruction targeted toward conceptual change is both necessary and beneficial for knowledge acquisition. Aulls describes how reciprocal teaching addresses the challenge of promoting conceptual change via dynamic teacher-student interaction.

Textbooks are dense composites. They lack narrative structure or personal touch compared with single-authored works; this distinction is shown in stage five. Armbruster's framing procedures help readers perceive organization in firmly packed text. Unpacking the material through charts helps students view relationships. Harrison illustrates a particularly apt frame: theory and evidence. It is easier to perceive and distinguish theory and evidence if they are recorded on charts.

Acquiring process skills in science education is fostered by writing, according to Santa and Havens. Writing permits and encourages originality. Through the written word a person may construct a unique hypothesis, generate an idiosyncratic interpretation, or form a personal idea. These ideas may be correct or incorrect; revised or kept intact; discarded or accepted. But they are unique to the person, which enhances their likelihood of originality. These tactics support the individualism and the juxtaposition of theory and evidence that are vital to scientific thinking.

The implication of the evolution of learning is simple: celebrate distinctions. Learning to read science texts and learning to think scientifically are not the same. Teachers can support students best by teaching the differences between scientific thinking and reading comprehension. Knowing how they differ, why they differ, and when to use the two types of cognitive strategies are essential skills. Learning to read text and learning to think like a scientist are different species from different families. But they will flourish if they are nourished in the same educational niche—the science unit.

References

Anderson, R., & Pearson, P.D. (1984). A schema-theoretic view of basic processes in reading. In P.D. Pearson (Ed.), *Handbook of reading research* (pp. 225-291). White Plains, NY: Longman.

Bronowski, J. (1973). *The ascent of man*. Boston, MA: Little, Brown.

Foucault, M. (1972). *The archeology of knowledge.* New York: Pantheon.

Graff, H.J. (1987). *The legacies of literacy.* Bloomington, IN: Indiana University Press.

Kuhn, D., Amsel, E., & O'Loughlin, M. (1988). *The development of scientific thinking skills.* San Diego, CA: Academic.

Popper, K.R. (1959). *The logic of scientific discovery.* New York: Harper & Row.

Pressley, M., & Ghatala, E. (1990). Self-regulated learning: Monitoring learning from text. *Educational Psychologist, 25*(1), 19-23.

Johnson, C., Symonds, S., McGoldrich, J., & Kurita, J. (in press). Strategies that improve children's memory and comprehension of what is read. *Elementary School Journal*

Sternberg, R. (Ed.). (1986). *Advances in the psychology of human intelligence.* Hillsdale, NJ: Erlbaum.

Author Index

Note: "Mc" is alphabetized as "Mac." An "f" following a page number indicates that the reference may be found in a figure.

Subject Index

Note: An "f" following a page number indicates that the reference may be found in a figure; a "t," that it may be found in a table.